International Political Economy Series

Series Editor

Timothy M. Shaw, University of Massachusetts Boston, Boston, USA;
Emeritus Professor, University of London, London, UK

The global political economy is in flux as a series of cumulative crises impacts its organization and governance. The IPE series has tracked its development in both analysis and structure over the last three decades. It has always had a concentration on the global South. Now the South increasingly challenges the North as the centre of development, also reflected in a growing number of submissions and publications on indebted Eurozone economies in Southern Europe. An indispensable resource for scholars and researchers, the series examines a variety of capitalisms and connections by focusing on emerging economies, companies and sectors, debates and policies. It informs diverse policy communities as the established trans-Atlantic North declines and 'the rest', especially the BRICS, rise. NOW INDEXED ON SCOPUS!

More information about this series at
http://www.palgrave.com/gp/series/13996

Eyene Okpanachi ·
Reeta Chowdhari Tremblay
Editors

The Political Economy of Natural Resource Funds

palgrave
macmillan

Editors
Eyene Okpanachi
International Centre for Policing and Security
University of South Wales
Pontypridd, UK

Reeta Chowdhari Tremblay
Department of Political Science
University of Victoria
Victoria, BC, Canada

ISSN 2662-2483 ISSN 2662-2491 (electronic)
International Political Economy Series
ISBN 978-3-030-78250-4 ISBN 978-3-030-78251-1 (eBook)
https://doi.org/10.1007/978-3-030-78251-1

This Palgrave Macmillan imprint is published by the registered company Springer Nature Switzerland AG
The registered company address is: Gewerbestrasse 11, 6330 Cham, Switzerland

Preface

This edited volume *The Political Economy of Natural Resource Funds* features contributions from an eclectic mix of scholars and professionals who are either engaged in different aspects of governance, politics, and development research in general, or who have specific expertise in natural resource management and sovereign wealth funds (SWFs). They, however, share the common goal of stimulating critical conversations and offering deeper empirical insights into the political economy of natural resource funds (NRFs) in different resource-rich countries throughout the world, including Angola, Botswana, Canada/USA, Chile, Ghana, Nigeria, Norway, Saudi Arabia, and Timor Leste.

Why does a specific focus on the politics of NRFs or SWFs financed predominantly from natural resource revenues matter? In this work, we have addressed a major lacuna in the current scholarly and policy literature by arguing that even though SWFs differ significantly in composition and in sources, most studies have not demonstrated sufficient awareness of such differences, or of the analytical and theoretical benefits of taking into account differences between the sources of these funds. This has hindered the development of a fine-grained understanding of how the sources of funding of SWFs may matter in the political dynamics of establishment and operations. By specifically focusing on NRFs or resource-based SWFs, this book aims to foster a better understanding of the particular challenges faced by resource-dependent countries or jurisdictions in managing their resource revenues. It also explores the varieties of natural resource management strategies as dictated primarily by domestic politics, and how the potential negative distributional consequences of resource wealth management (the resource curse) or efforts to combat it may add political dimensions and potential conflicts to decisions about NRFs in ways that other SWFs do not experience.

In explicating how NRFs as initiatives that are, or are perceived as, capable of (re)configuring power relationships between governments at different levels and between the state and society, we hope that this edited volume might provide useful insights into two sets of questions relating to: (a) power relations in the larger governance and management of natural resource wealth; (b) the concept of time. As institutions, NRFs are influenced by historical processes such as exogenous shocks or fluctuations in commodity prices or the emergence of new leaders/rulers. Periods such as these may unravel, or hasten the unraveling of, the "consensus" over NRFs, increase the mobilization of groups or actors concerning NRFs, and may lead to shifts or promote continuity in the design or operation of NRFs. The book, thus, makes an attempt to highlight the dynamic nature of the politics of NRFs, taking into consideration not just how the politics is shaped by historical changes or contingency events, but also how NRFs are shaped by the interests of political or economic actors and their shifting political coalitions. Moreover, by paying close attention to the players and their positions, processes and mechanisms, and political-economic outcomes of NRFs, this volume illustrates the similarities and differences in resource revenues management in countries whose political, economic, and social circumstances, including the degree of dependence on resources, vary greatly, and underscores how these circumstances have changed over time and the effects of these changes on NRFs. By bridging the existing academic and practical knowledge gap arising from the limited attention given to the domestic politics of NRFs and state–society relations, we hope this edited collection will serve as a valuable resource for academics, policymakers, and civil society actors in resource-driven economies, and especially those interested in learning from comparative experiences of natural resource wealth management through NRFs.

This book is the result of Eyene's Government of Canada funded Banting Postdoctoral Fellowship project which he completed under Reeta Tremblay's supervision at the Department of Political Science, University of Victoria (UVic), Canada, 2017–2019. Eyene would like to extend special thanks to Amy Poteete (Concordia University) for her valuable suggestions during the Banting Fellowship application process. We owe a great gratitude to Xu Yi-chong for her invaluable comments and suggestions which helped us refine our thoughts on the book project. We gratefully acknowledge the generous funding of the Social Sciences and Humanities Research Council of Canada (SSHRC) for its support Connection Grant 2019–2020 which facilitated the hosting of the scholarly workshop in Victoria in September 2019 that brought together the authors to present their draft chapters for this edited volume. We would also like to thank the Faculty of Social Sciences UVic; the Centre for Global Studies (CFGS) UVic; and the Centre for Governance and Public Policy, School of Government and International Relations, Griffith University, Australia, for their financial support of our scholarly workshop. We would like to thank our authors for contributing their scholarship, for sparing time from their busy schedule to attend the Victoria workshop, and for their commitment and patience in bringing the book to fruition from the development of the project idea through the review and editorial phases. We are also grateful to the Department of Political Science, UVic, for providing a conducive environment and much-needed institutional support. We are indebted to the department's staff, particularly Rosemary Barlow and Joanne Denton, for their great help in planning and coordinating the workshop logistics, and in administering the finances for the workshop. Thanks also to Jodie Walsh of the CFGS for her support during the workshop, and to Holland Gidney, Research and Scholarship Coordinator for the Faculty of Social Sciences, and Nicole Kitson of the Office of Research for their feedback on our Connection Grant application. Our thanks also go out to the two doctoral students Smith Oduro-Marfo and Nicole Bates-Eamer, who have assisted us with, respectively, conference organization and editorial work. Our thanks also go to Cindy Chopoidalo for proofreading the manuscript. This project would not have been possible without the support of Tim Shaw, series editor of Palgrave Macmillan's IPE. His constant encouragement pushed us to complete this manuscript in a timely manner. We are grateful to the editorial team at Palgrave Macmillan, and Anca Pusca, Executive Editor, International Studies, for their support. Finally, Eyene would like to express his deepest gratitude to his spouse, Esther, and their children, Ivanna and

Jaden, for their patience, understanding, and support, without which this would not have been possible.

Victoria, Canada

Eyene Okpanachi
Reeta Chowdhari Tremblay

The original version of this book was revised: The correct name Joanne Denton was updated in the Preface. The correction to this book is available at https://doi.org/10.1007/978-3-030-78251-1_12

Contents

Notes on Contributors

Ishmael Ackah is an Energy Economist and a Fellow of the Institute for Economic Affairs Ghana. He has over 10 years' experience that spans public service, civil society, private sector, and academia. Currently, he is a Lecturer with Ghana Institute of Management and Public Administration, and the Regulatory and Electricity Market Expert of the USAID West Africa Energy Program. He worked as the Head of Policy Unit at the Africa Centre for Energy Policy, served as the Technical Advisor on Energy and Petroleum Policies to Ghana's Minister for Planning and was the first Coordinator of Local Content Secretariat at the Ghana Energy Commission. He has provided research consultancy services for some local and international institutions such as the United Nations University, the African Development Bank, IHS Markit, SNV Ghana. Among his publications are 40 peer-reviewed papers in high-impact journals. Ishmael holds a Ph.D. in Energy Economics and Policy from the University of Portsmouth, UK and an MSc also in Energy Economics and Policy, from the University of Surrey, UK.

Terhemba N. Ambe-Uva is a Ph.D. researcher in political studies at the University of Ottawa, Canada, and a research fellow at the West Africa Institute of regional integration and social transformation, Praia, Cape Verde. His doctoral work in global ecological political economy examines the power and material interests underpinning the governance of the sustainable fishmeal industry in West Africa. Terhemba has written on comparative regional integration as a team member of a University of Bonn project on "Sustainable Regional Integration in West Africa and Europe" and has published on diaspora policies, mining reforms, and just energy transitions.

Sara Bazoobandi is a Marie Curie Fellow with the Institute of Middle East Studies at the German Institute of Global and Area Studies. She is also a Non-resident Senior Fellow at the Atlantic Council, and a Non-resident Fellow at the Arab Gulf States Institute in Washington. Among her publications is *The Political Economy of the Gulf Sovereign Wealth Funds* published by Routledge.

Reeta Chowdhari Tremblay is a Professor of Comparative Politics and former Provost/Vice-President Academic at the University of Victoria, Canada. Her recent publications include *Religion and Politics in Jammu and Kashmir* (2020), *CoVid India: Federalism, Majoritarian Nationalism, and the Vulnerable and Marginalized* (2020), and *Modi's Foreign Policy* (2017). She is also the leading author of the landmark book, *Mapping the Political Landscape: An Introduction to Political Science*, which is an authoritative student guide on the science and methods of studying politics. She is frequently sought after for commentaries and opinions on diverse range of issues relating to governance, politics, and public administration, especially in the context of the Global South.

Jerry Courvisanos is an Associate Professor of Innovation and Entrepreneurship, Federation Business School. Previously, he has held the roles of Associate Dean of Research and Director of the Centre for Regional Innovation and Competitiveness. He is currently the Timor-Leste Relationship Coordinator of Federation University Australia and Director of the International Co-operative Centre with Dalian University of Technology. Jerry's basic research is centred on understanding the processes of innovation and their impact on investment spending, business cycles, sustainable development, and the long-term development of businesses and the economy. He has published many articles, book chapters, and books on these topics.

Anita Doraisami currently teaches Economics at Federation University, Australia, and previously taught Economics at Monash University in Melbourne, Australia. She has served as a consultant for various organizations including the Asian Development Bank, USAID, UNRISD, and SIDA. In addition, she was a long-term consultant to the International Monetary Fund in Singapore and Macroeconomic Adviser to the Ministry of Finance, Timor Leste. She holds a Ph.D. in economics from the

University of Cambridge and has published articles on the management of Timor-Leste's natural resources.

Denis M. Gyeyir is currently the Africa Program Officer with the Natural Resource Governance Institute (NRGI). At NRGI, Denis leads on data use, revenue management, SOE governance, and energy transition work. Prior to joining NRGI, he was a Technical Officer and later Senior Technical Officer and Assistant Technical Manager at the Secretariat of the Public Interest and Accountability Committee (PIAC), Ghana. At PIAC, Denis played a key role in compiling the PIAC reports on the management of petroleum revenues in Ghana, monitoring compliance and ensuring that petroleum revenues are managed in accordance with law. Denis is a researcher and certified reviewer and has written about various aspects of the extractive sector in Ghana including revenue management, commodity trading, management of sovereign wealth funds, energy transition among others. He has published in the *Journal of Energy Policy and the International Journal of Economics and Financial Research.*

Diego Leiva has a Ph.D. from the School of Government and International Relations, Griffith University. His work focuses on Latin American presidentialism and foreign policy, Chinese investment in Latin America, and Sino–Latin American relations. His latest publications are "BRI and railways in Latin America: how important are domestic politics?" In Asian Education and Development Studies (2019), and "Xi Jinping and Sino–Latin American Relations in the 21st Century: Facing the Beginning of a New Phase?" in *Journal of China and International Relations* (2017).

Sarah J. Martin is an Assistant Professor in the Department of Political Science at Memorial University of Newfoundland and Labrador, where she researches the dynamics of food, feed, and fuel in relation to environmental politics. She co-edited (with Ryan Katz-Rosene) Green Meat: Sustaining Animals, People and the Planet (McGill-Queen's University Press, 2020).

Jonathon W. Moses is a Professor in the Department of Sociology and Political Science at the Norwegian University of Science and Technology (NTNU) Trondheim, Norway. He is the author (with Bjorn Letnes) of *Managing Resource Abundance and Wealth: The Norwegian Experience* (2017, Oxford University Press).

Eyene Okpanachi is a Marie Curie Fellow at the International Centre for Policing and Security (ICPS), University of South Wales. He was previously Banting Postdoctoral Fellow at the University of Victoria and Vanier Canada Graduate Scholar at the University of Alberta. His Banting postdoctoral research from which this book project emerged focused on the institutions and politics of Natural Resource Funds (NRFs). His works on natural resources, political institutions, and conflict processes have been published in journals such as *Review of Policy Research*, *Publius: The Journal of Federalism*, World Futures, and *Commonwealth and Comparative Politics*.

David Sebudubudu is a Professor of Political Science, and Dean of the Faculty of Social Sciences, University of Botswana. He is an editor of the *Africa Yearbook: Politics, Economy and Society South of the Sahara*, published by Brill (Leiden & Boston), from January 2020.

Peter J. Smith is a Professor Emeritus of Political Science at Athabasca University, Alberta, Canada. He has written on the political economy of sub-federal governments in Alaska and Alberta. His current research interests include new communications technologies, globalization, religion, trade politics, transnational networks, democracy, and citizenship. He is the co-editor of *Religious Activism in the Global Economy: Promoting, Reforming or Resisting Neoliberal Globalization* (2016) with Sabine Dreher, and co-editor (with Katharina Glaab, Claudia Baumgart-Ochse, and Elizabeth Smythe) of *The Role of Religion in Struggles for Global Justice* (2018).

Xu Yi-chong Professor, Center for Politics and Public Policy, Griffith University, Australia, is the author of *The Sinews of Power* (2017, OUP), *The Politics of Nuclear Energy in China* (2010); *Electricity Reform in China, India and Russia* (2004); co-author of *International Organisations and Small States* (with J Corbett, P Weller 2021), *The Working World of International Organisations* (with P. Weller, 2018), and co-editor (with G. Bahgat) *The Political Economy of Sovereign Wealth Funds* (2010). All the projects were supported by research grants from the Social Sciences and Humanities Research Council of Canada (SSHRC) and Australian Research Council (ARC).

List of Figures

To Save or Not to Save: State Governments and the Construction of Privilege over the Creation of Oil Funds in Nigeria

A Less Than Sovereign Wealth Fund: Norway's Government Pension Fund, Global

Old Fund, New Mandate: Saudi Arabia's Public Investment Fund (PIF)

Practices and Models: Prospects for Commodity-Based SWFs

List of Tables

Saving Today's Bread for Tomorrow's Consumption? The Politics of Trade-offs in the Governance of Ghana's Petroleum Funds

To Save or Not to Save: State Governments and the Construction of Privilege over the Creation of Oil Funds in Nigeria

A Less Than Sovereign Wealth Fund: Norway's Government Pension Fund, Global

Old Fund, New Mandate: Saudi Arabia's Public Investment Fund (PIF)

The Timor-Leste Petroleum Fund: From Buying Peace to White Elephants

Practices and Models: Prospects for Commodity-Based SWFs

Introduction: The Political Economy of Natural Resource Funds (NRFs)

Eyene Okpanachi and Reeta Chowdhari Tremblay

The puzzling phenomenon of resource-abundant countries that, contrary to expectations, perform worse than resource-poor countries on most development indicators has been receiving significant attention in the political economy literature, especially since the seminal work of Richard Auty (Auty 1993; Sachs and Warner 1995; Karl 1997; Ross 1999). One important mechanism in this negative relationship is the mismanagement of resource revenues, especially during periods of booms (Ashafa 2007; Kaufmann and Sachs 2014), which leaves resource-dependent countries and subnational governments vulnerable to the damaging effects of resource revenue volatility. Some of the mechanisms by which poor management of windfall revenues contributes to the resource curse include expansion of current account expenditure, investment in ill-conceived and ambitious but low-return projects, accumulation of foreign debts, outright theft of resource wealth by political elites, or lack of

---

E. Okpanachi (✉)
University of South Wales, Pontypridd, UK
e-mail: eyene.okpanachi@southwales.ac.uk

E. Okpanachi · R. Chowdhari Tremblay
University of Victoria, Victoria, BC, Canada
e-mail: reeta@uvic.ca

E. Okpanachi and R. Chowdhari Tremblay (eds.), *The Political Economy of Natural Resource Funds*, International Political Economy Series,
https://doi.org/10.1007/978-3-030-78251-1_1

incentives to build strong institutions (Ashafa 2007; Collier and Venables 2011).

Natural Resource Funds (NRFs) or resource-based Sovereign Wealth Funds (SWFs)[1] are considered as important institutional mechanisms to mitigate or ameliorate the resource curse (van der Ploeg 2008; Frynas 2017; Alsweilem and Rietveld 2017; Oteh 2017). De-linking government expenditures from volatile resource revenues by saving or investing windfalls in NRFs during prosperous times, and drawing down those funds during downturns, lowers the negative consequences of boom-and-bust cycles on macroeconomic performance and human development (Ashafa 2007; ODI 2006). Ring-fencing resource revenues also help to shield these revenues from corruption and mismanagement (Bauer 2014). In addition, saving or investing these funds also promotes intergenerational equity, which is important because natural resources are exhaustible (Lipschitz 2011). Indeed, NRFs are viewed not only as institutional responses to the resource curse, but also as potential "useful step[s] towards the stimulation of a solid and durable institutional context" or conditions for economic growth in resource-rich countries in general (Baena et al. 2012, p. 571).

However, although many countries have established NRFs, the effectiveness of NRFs varies (Davis et al. 2001; Humphreys and Sandbu 2007; Kaufmann and Sachs 2014). In some countries, NRFs have reined in profligate spending, whereas in others, NRFs have been mismanaged (Davis et al. 2001; Humphreys and Sandbu 2007), and some funds have not met their establishment mandates. For instance, many funds set up to perform stabilization roles during periods of declining revenues, such as those of Azerbaijan, Kazakhstan, Trinidad and Tobago, and Venezuela, have failed to stabilize their budgets (Kaufmann and Sachs 2014, p. 4). In other cases, funds preserved in NRFs are raided by governing elites (Humphreys and Sandbu 2007). These mixed results have led Davis et al. (2001) to describe oil-based NRFs as "problems posing as solutions," suggesting that from their econometric study, overall, these "funds may work only when they are not needed." Although the note of caution about the expectations of NRFs is useful, it is also important to note that "for

[1] The Sovereign Wealth Fund Institute (SWFI) [2016] defines an SWF as "a state-owned investment fund or entity that is commonly established from balance of payments surpluses, official foreign currency operations, the proceeds of privatizations, government transfer payments, fiscal surpluses, and/or receipt resulting from resource exports."

fiscal policy to be correct, it is not sufficient to get the economics right"; explicit attention must also be given to the question about how politics affects economic processes (Humphreys and Sandbu 2007, p. 199). This is more so because as pointed out by Karl (2007, p. 256), "the 'resource curse' is primarily a political and not an economic phenomenon."

How does politics influence the establishment and operation of NRFs? Most existing studies have tended to view or examine the political and economic behavior of SWFs on the international scene, or the countries hosting SWFs' foreign investment (Truman 2010; Yi-chong 2010). Commenting on this excessive focus on the external orientation of SWFs, for example, Cummine (2016, p. 3) noted the lack of attention in the Santiago Principle, the 24 sets of guiding principles and practices that over 30 SWFs have voluntarily agreed to respect, to the fundamental issue related to the nexus between these funds and domestic citizen–state relationship. Also, conventional scholarship has, for the most part, viewed NRFs as subsets of SWFs. This characterization is appropriate, although it has created analytical and theoretical problems for the study of resource-based SWFs or NRFs, which make up 57% of SWF assets (Grigoryan 2016). A major issue is a tendency in a significant proportion of literature to view NRFs, like other types of SWFs, largely as financial or investment vehicles that enhance the sponsoring states' economic geopolitical interests or strategic power internationally. The primary question in these literatures is whether these Funds' parent countries are using them as "instruments of state power" (Balding 2012, p. 72) for geopolitical interest (Griffith-Jones and Ocampo 2011; Helleiner 2009) or the preservation of autonomy and sovereignty (Monk 2011). Accordingly, much less attention has been given to analyses of the domestic political economy drivers of these Funds' behavior or the crucial roles of the constitutive sources of SWFs in shaping the politics surrounding them and, by implications, their establishment, behavior, and performance.

A growing body of literature explicitly focuses on the domestic political ramifications of SWFs (for example, Shih 2009; Bahgat 2010; Pistor and Hatton 2011; Balding 2012; Clark et al. 2013; Grigoryan 2016). For instance, Shih (2009) attributed the divergent performances of China's and Singapore's SWFs to the degree of domestic unity each country enjoys. He argued that the tightly coupled nexus between Singapore's political elites and business interests created a stable political climate that made the SWF successful, whereas the fragmented nature of this relationship in China created bureaucratic infighting that undermined success.

Balding (2012, pp. 77–78) revealed how SWFs are imbricated in domestic state capitalism, and concluded, using "a detailed analysis of domestic political interactions between the government and the SWFs," that "SWFs [have] become political creatures within the domestic framework that fights for resources and influence like numerous other actors and institutions to which the government responds." These are indeed important insights. However, the domestic political economy of SWFs has largely been examined in relation to non-resource-based SWFs, the majority of the exceptions being work on oil-based Gulf SWFs, such as Bahgat's case study of Kuwait (2010), Bazoobandi's study of Iran, Kuwait, Saudi Arabia, and the United Arab Emirates (2013), Grigoryan's study of Abu Dhabi (2016), and Braunstein's study of Kuwait, Abu Dhabi, Qatar, and Bahrain (2019).

Our collection gathers some of the most comprehensive and holistic analyses of case studies of NRFs to date, explicitly focusing on the interaction between political economy and NRFs in those Funds' parent countries in order to produce rich and detailed knowledge of the dynamics of NRFs, including their external behavior. It complements extant contributions on NRFs such as Bacon and Tordo's exploration of the institutional and financial performance of oil funds (2006), Humphreys and Sandbu's (2007, p. 195) theoretical expose on the optimal institutional design or "fixes" of NRFs, Bauer's surveys of the laws, regulations, and policies governing NRFs (2014), Alsweilem and Rietveld's analysis of the "institutional and fiscal foundations" of these funds (2017), Frynas's expose on resource funds and governance (2017), and Al-Hassan and colleagues' examination of the macro-fiscal linkages for stabilization and savings funds (2018). It however builds on these contributions through the explanatory primacy it accords the political contestations, bargaining, and compromise (or lack of it) over these Funds.

The contributors to this book focus on how domestic political struggles and contestations in these countries shape NRF decision-making processes, such as establishing, saving in, investing in, and governing these Funds. The contested politics of NRFs may include, for instance, those between different levels and tiers of government, between governing elites or factions of ruling monarchy for political power, between different arms of government (legislature, executive, and judiciary), bureaucratic competition, partisan disagreements, business interests' opposition or support, or civil society's contestations.

By specifically focusing on NRFs or resource-based SWFs, this volume does more than merely examine the particular challenges faced by resource-rich countries or jurisdictions. It also helps us better illuminate the varieties of natural resource management strategies as dictated primarily by local politics, and how the potential negative distributional consequences of resource wealth management (or the resource curse) may make decisions regarding NRFs decidedly political and contentious in ways that other forms of SWFs may not be. In turn, the latter insights are important contributions to the debate over the resource curse and the questions of whether or not, and under what political conditions, resource funds can help in mitigating this "curse." These insights help us further understand the politics of resource-dependent economies, the political, social, and economic forces that shape resource wealth management, and the continuity and changes of these interactions. The central emphasis here is a more focused investigation of domestic politics and the political forces that shaped decisions made about the funds, bearing in mind the basic understanding of politic as a highly contested process (Lasswell 1936).

Thus, our two central research questions in this volume are: What are the domestic political forces that have shaped decision-making over the creation and operation of NRFs; and what are the implications of the struggles over this issue for political power, material interest, and legitimacy? Like the management of natural resource revenues itself, NRFs may be challenged or contested due to perceived crises of legitimacy, accountability, and transparency, with the possibility of heightened contestations in decisions concerning NRFs between different stakeholders within resource-rich countries, such as elites representing the ruling coalition and governments, and between society and the state. Legitimacy is a central concern in any decision-making process, including the establishment and operation of SWFs (Clark et al. 2013; Rose 2019). This question of legitimacy, we argue, is more pronounced with NRFs or resource-based SWFs than other sources of SWFs. This is not just because of the exhaustive nature of these resources (Ang 2012), which makes the question of reallocation of resource wealth for the future all the more salient, but also because of the damaging effects of resource revenue on institutional quality. This is especially true for countries with ineffective institutions (Mehlum et al. 2006), which in turn can negatively affect the sustainable institutionalization of NRFs as oil windfalls are consumed rather than saved. This latter point also calls to attention the link between resource windfalls and rentierism (Beblawi and Luciani 1987; Beblawi 1987b; Herbst 2000) which, depending on context, can either undermine accountability or exacerbate struggles over resource rents, or both.[2] For Beblawi (1987a), a rentier economy, from which a rentier state develops, is one in which economic rents paid by foreign companies accrue directly to the government, and one in which a small proportion of the population is engaged in rent generation. Free from the responsibility of taxation because of the easy inflow of large resource rents, which is also used to undermine the formation of an active citizenry capable of demanding accountability from government (Ross 2001) rentier states are able to operate largely unconstrained "by the type of *internal countervailing pressures* that helped to produce bureaucratically efficacious, authoritative, liberal, and ultimately democratic states elsewhere" (Karl 2007, p. 262). The rentier state not only impedes the emergence of independent social classes capable of advocating for open government; it also negatively affects political institutions. In addition, the zero-sum struggles over the capture of oil rents by elites intensifies conflict, undermines democracy, and entrenches authoritarianism, as the elites use resource funds to suppress independent groups that would otherwise pressure them for reform (Busse and Gröning 2013; Ross 2015). Unencumbered by significant pressures for accountability from social groups, the elites are thus free to misuse their power, which fuels corruption and further weakens institutional quality (Busse and Gröning 2013; Ross 2015).

In addition to its vulnerability to rentierism (see Ambe-Uva & Martin, in this volume), which can undermine the whole idea of a NRF, resource wealth is also subject to uncertainty over whether the state or society, or both, own that wealth, which makes the sustainability of NRFs very difficult to attain. Similarly, NRFs embody conflicting objectives, which

[2] We acknowledge, as scholars writing on the changing political economy of rentier state or "late" rentierism (for example, Gray 2011; Yamada and Herto 2020) have done, that rentierism is not predestined, is open to shifts amid continuity, and states' rentier practices can be challenged from below, even if partial. As Bazoobandi noted (in this volume), a confluence of factors, including the oil price collapse of 2014, the emergence of crown prince Mohammed bin Salman as likely heir to the throne, and the launch of the economic and social reform program Vision 2030, strained the traditional 'rentier bargain' in Saudi Arabia. These factors have led to a change of rentier strategies regarding the distribution and sharing of oil rents, with the Public Investment Fund (PIF) playing a central role in this changing rentier dynamics and continuing (re)negotiations of power balance between the ruling family and between state and society.

can cause a divergence between the interests of supporters and opponents. One such example is the tension between the normative objective of saving resource windfalls for future generations and pressures to spend these windfalls to solve immediate government needs. Consensus can emerge over these questions, as the Norwegian case has largely demonstrated (see Moses, in this volume); however, even in this case, such a consensus is far from automatic and is, inherently, a political process. The case studies in this volume are therefore interested in explaining the political processes that lead to such a consensus. This analysis becomes more important because, even when such an alignment of interests exists, those interests can be ruptured and shaken during periods of natural resource price change, whether downturns or increases. Such periods may bring into the open the subdued consensus between different branches or tiers of government, factions of a ruling family, political parties, labor, resource-bearing communities, and civil society groups over the question whether or not to save resource windfalls and to what extent.

In explicating NRFs as initiatives or policy responses that are, or are perceived as, capable of reconfiguring power relationships—between governments, ruling elites, political parties, bureaucrats, and between the state and society—and are hence deeply contested, we hope to provide insights into the question of power relations in the larger governance and management of natural resource wealth. In doing so, this collection provides a critical assessment/analysis of NRFs in a way that gives serious attention to the concept of temporality or time. As institutions, NRFs are influenced by historical processes such as exogenous shocks or fluctuations in commodity prices. The high volatility of resource wealth is central to the issue of temporality. As Humphreys et al. (2007, p. 6) explain, "the volatility of income comes from three main sources: the variation over time in rates of extraction, the variability in the timing of payments by corporations to states, and fluctuations in the value of the natural resource produced." Periods such as this may unravel, or hasten the unraveling of, the "consensus" over NRFs between factions of ruling elites, or undermine the distributive prowess of the state and hence provide opportunities for the recalibration of state-society relations. The mobilization of groups or actors concerned about NRFs may thus increase, which may ultimately heighten contentions and lead to shifts or continuity in the design or operation of NRFs. This illuminates the dynamic nature of the politics of NRFs, taking into consideration the influence not only of changes in commodity prices on the politics, but also of institutional formation and change, especially how NRFs are shaped by historically contingent events or shifting political alliances. This latter insight is important because, as Poteete points out, institutions are themselves "products of politics" and are hence "subject to change with political conditions" (2009, p. 545). Accordingly, our objectives in this anthology are two-fold. The first is to provide a better understanding of the role of political agency—of state and nonstate actors—in the decisions to credibly commit to policies such as NRF establishment. The second is to unravel the distributional tensions and contestations that take place over the operation of these policies and institutions over time, particularly at such important historical times as periods of new resource discovery, resource price booms or downswings, electoral cycles and changes of government/ruling elites, and periods of changing political coalition, all of which contribute to intensified conflict over NRFs, including pressure for new institutional rules or changes to existing ones. We believe that attention to these temporal issues is essential to explicate the political economic dynamics underlying and shaping NRFs.

This volume evaluates ten NRFs, with divergent historical, political, and economic experiences, from countries or subnational jurisdictions in North America, Western Europe, Sub-Saharan Africa, Southeast Asia, Latin America, and the Middle East. These diverse countries were chosen to illuminate the common challenges they confront and the practices they have devised in establishing and operating their NRFs, as well as their differences in outcome based on context, including political regimes, political cultures, and the state–society complex (Table 1).

In order to ensure a consistent explanation of the domestic political ramifications of NRFs, we asked our contributors to think about generating, to the extent possible, an interlinked three-stage analysis: Pre-establishment stage; once established, how NRF should be governed (i.e., governance and management framework of the funds); and finally NRF in operation. We asked them specific questions to address in each stage of analysis, depending on the country context. For example, in the Pre-establishment stage, we asked them to review the key elements of the debate about whether or not to establish a fund; what the proposals were, who proposed what and why, and the different players' interpretations of these proposals. We also asked them to identify the salient factors or configuration of factors central to the social and political formations in their country of focus that shaped the debates or "fiscal politics" over the establishment of the fund and the conflict and compromise that

Table 1 Brief information of the ten NRFs covered in this volume

Fund name	*Assets under management(US$ billion, 2021)*[i]	*Country*	*Year established*	*Funding source*
Government Pension Fund Global	1,289.46	Norway	1990	Petroleum
Public Investment Fund	430.00	Saudi Arabia	1971	Petroleum
Alaska Permanent Fund	67.34	USA	1976	Petroleum
Timor-Leste Petroleum Fund	18.46	Timore-Leste	2005	Petroleum
Alberta Heritage Savings Trust Fund	17.80	Canada	1976	Petroleum
Chile Funds (Fondo de Estabilisación Económica y Social or ESSF and the Pensions Reserve Fund Fondo de Reserva de Pensiones, or PRF)	15.71	Chile	2007	Minerals
Pula Fund	4.92	Botswana	1994	Minerals
Fundo Soberano de Angola	3.24	Angola	2012	Petroleum
Nigeria Sovereign Investment Authority	2.56	Nigeria	2011	Petroleum
Ghana Petroleum Funds (Ghana and Stabilisation Funds)	0.97	Ghana	2011	Petroleum

Source Authors' compilation based on data from Sovereign Wealth Fund Institute (2021). The assets value of the Alberta Heritage Savings Fund was sourced from Government of Alberta (2021)

[i] The figures cited by individual contributors may be slightly different given that the ones provided here are the latest

eventually produced a particular type of NRF. For instance, they were asked to illuminate whether and how these debates were affected by electoral cycles, vertical division of governmental power (federal vs unitary states), horizontal division of governmental power (executive-legislature-judiciary, finance or national planning ministry, central bank, technical committees, government's economic team, etc.), ideological orientation of governing parties (left vs right); regime type (authoritarian or democratic), organized interests such as citizens groups, business groups, or extractive industry's representatives; and regional geopolitical politics, etc. In the establishment stage, the contributors were asked to give attention to the governance regime, such as the legal framework, institutional framework and governance structure, formal rules and informal norms, investment and risk management framework. Each contributor was asked to address the management of the fund and its actual operationalization by examining questions such as who should manage the fund (Finance minister", central bank, or an arms-length agency)? Whether the NRFs had a board and management team? Who decides on investment—the bureaucrats or the politicians? And where to invest—domestic vs external, rich vs poor. Finally, at the NRFs in operation stage, we asked the contributors to the patterns of political contestation that emerged over the governance/management arrangements of the funds as set out in stage two above, the conflicts over the actual operation of the fund or the unresolved issues in stage one, the processes of conflict mitigation, and the changes (if any) made to the funds as a result of the conflict over its establishment or operation. Also, we are interested in the political dynamics underlying the NRF, the contributors were asked to be attentive to continuity and changes in the funds operations and to highlight the historical junctures around which political tensions over the funds or pressures for change or reforms of the funds are loudest or more pronounced—is it during election, during resource revenue boom or price downturn, change in governing regime?

The theme of this collection is relevant for a variety of issues and subjects at the intersection of sustainable use of natural resource revenues: public finance and economics, law, international relations, policy and innovation, and politics of decision-making. We believe that outside the scholarly community, practitioners in the fields of natural resource management, academics, governments, and civil society groups will find this work helpful. The relevance of the book for policymakers and practitioners emanates from its potential to bridge the existing academic and practical knowledge gap arising from the limited attention given to domestic political economy in most extant literature on SWFs. The relatively low availability of such knowledge has undermined not only the critical task of identifying workable reforms, but also, and more importantly, the key constituencies that drive or block reforms. With many resource-rich countries or subnational jurisdictions establishing NRFs (Kaufmann and Sachs 2014), the need for such empirical knowledge that could inform policy is more urgent than before. The renewed attention to NRFs due to the recent volatility of oil prices either due to rising supply glut or the coronavirus pandemic (COVID-19) and the negative effects of such volatility on the political economies of oil-rich countries makes this knowledge all the more relevant. It also draws attention to the potential role of NRFs in mitigating adverse effects of oil price downturns. We hope that this collection, given its timely nature, will be an indispensable compendium to policymakers looking for best ways to manage their countries' natural resource revenues to avoid the resource curse, while providing evidence-based case study analyses to the scholarly community and to the civil society organizations who advocate for transparency, accountability, and fiscal responsibility in the use of natural resource revenues.

References

Al-Hassan A et al. (2018) Commodity-based sovereign wealth funds: Managing financial flows in the context of the sovereign balance sheet. International Monetary Fund Paper, WP/18/26

Ang A (2012) Four benchmarks of sovereign wealth funds. In: Bolton P, Samama F, Stiglitz J (eds) Sovereign wealth funds and long-term investing. Columbia UP, New York, p 94–105

Ashafa S (2007) National revenue funds: Their efficacy for fiscal stability and inter-generational equity. *IISD*. www.iisd.org/pdf/2007/trade_price_nat_rev_funds.pdf

Alsweilem K, Rietveld M (2017) Sovereign wealth funds in resource economies: Institutional and fiscal foundations. Columbia UP, New York

Auty R (1993) Sustaining development in mineral economies: The resource curse thesis. Routledge, London

Bacon R, Tordo S (2006) Experiences with oil funds: Institutional and financial aspects. International Bank for

Reconstruction and Development/The World Bank. worldbank.org/curated/en/952141468141870800/pdf/ESM3210Experiences1Oil1Fund01PUBLIC1.pdf
Baena C, Sévi B, Warrack A (2012) Funds from non-renewable energy resources: Policy lessons from Alaska and Alberta. Energy Policy 51:569–577
Balding C (2012) Sovereign wealth funds: The new intersection of money and politics. Oxford UP, New York
Bauer A (ed) (2014) Managing the public trust: How to make natural resource funds work for citizens. Natural Resource Governance Institute & Columbia Center on Sustainable Development, New York
Bazoobandi S (2013) Political economy of the Gulf Sovereign Wealth Funds: A case study of Iran, Kuwait, Saudi Arabia and the United Arab Emirates. Routledge, London and New York
Bahgat C (2010) Kuwait Investment Authority-An assesment. In: Xu Y, Bahgat G (eds) The political economy of sovereign wealth funds. Palgrave Macmillan, Basingstoke, Hampshire, p. 72–87
Beblawi H (1987a) The Rentier state in the Arab World. Arab Studies Quarterly 9(4):383–398
Beblawi H (1987b) The rentier state in the Arab World. In: Beblawi H, Luciani G (eds) The rentier state. Abingdon, Oxon: Routledge, p. 49–62
Beblawi H, Luciani G (1987) Introduction. In: Beblawi H, Luciani G (eds) The rentier state. Abingdon, Oxon: Routledge, p.1–21
Braunstein J (2019) Domestic sources of twenty-first-century geopolitics: Domestic politics and sovereign wealth funds in GCC economies. *New Political Economy* 24(2):197–217
Busse M, Gröning S (2013) The resource curse revisited: Governance and natural resources. Public Choice 154(1–2):1–20
Clark G, Dixon A, Monk A (2013) Sovereign wealth funds: Legitimacy, governance, and global power. Princeton UP, Princeton
Collier P, Venables AJ (eds) (2011) Plundered nations: Successes and failures in natural resource extraction. Palgrave, London
Cumming D, Filatotchev I, Reinecke J, Wood G (2017) Introducing sovereign wealth funds. In: Cumming D, Wood G, Filatotchev I, Reinecke J (eds) The Oxford handbook of sovereign wealth funds. Oxford UP, Oxford, p 3–15
Cummine A (2016) Citizens' wealth: Why (and How) Sovereign Funds should be managed by the people for the people. Yale University Press, New Haven and London
Davis J et al. (2001) Oil funds: Problems posing as solutions? Finance and Development 38(4), http://www.imf.org/external/pubs/ft/fandd/2001/12/davis.htm.

Frynas G (2017) Sovereign wealth funds and the resource curse: Resource funds and governance in resource-rich countries. In: Cumming D, Wood G, Filatotchev I, Reinecke J (eds) The Oxford handbook of sovereign wealth funds. Oxford UP, Oxford, p 123–144

Government of Alberta (2021) Heritage Savings Trust Fund. https://www.alberta.ca/heritage-savings-trust-fund.aspx

Gray M (2011) A theory of "late rentierism" in the Arab States of the Gulf. Occasional Paper no. 7, Centre for International and Regional Studies, Georgetown University School of Foreign Service in Qatar. https://repository.library.georgetown.edu/bitstream/handle/10822/558291/CIRSOccasionalPaper7MatthewGray2011.pdf

Griffith-Jones S, Ocampo JA (2011) The rationale for sovereign wealth funds from a development perspective. In: Bolton P, Samama F, Stiglitz JE (eds) Sovereign wealth funds and long-term investing. Columbia UP, New York, p 60–66

Grigoryan A (2016) On the political economy of sovereign wealth funds. Dissertation, Universitat Siegen. dokumentix.ub.uni-siegen.de/opus/volltexte/2016/1059/pdf/Dissertation_Grigoryan_Artur.pdf

Helleiner E (2009) The geopolitics of sovereign wealth funds: An introduction. Geopolitics 14(2): 300–304

Herbst J (2000) States and power in Africa: Comparative lessons in authority and control. Princeton University Press, Princeton

Humphreys M, Sachs J, Stiglitz J (2007) Introduction: What is the problem with natural resource wealth? In: Humphreys M, Sachs J, Stiglitz J (eds) Escaping the resource curse. Columbia University Press, New York, 1–20

Humphreys M, Sandbu M (2007) The political economy of natural resource funds. In: Humphreys M, Sachs J, Stiglitz J (eds) Escaping the resource curse. Columbia UP, New York, p 194–233

Karl T (1997) The paradox of plenty: Oil booms and petro states. U of California P, Berkeley and Los Angeles

Karl T (2007) Ensuring fairness: The case for a transparent fiscal social contract. In: Humphreys M, Sachs J, Stiglitz J (eds) Escaping the resource curse. Columbia UP, New York, p 256–285

Kaufmann D, Sachs L (2014) Foreword. In: Bauer A (ed) Managing the public trust: How to make natural resource funds work for citizens. Natural Resource Governance Institute & Columbia Center on Sustainable Development, New York, p 1–2

Lasswell H (1936) Politics: Who gets what, when, how. Whittlesey House, New York

Lipschitz L (2011) Overview. In: Areki R, Gylfason T, Sy A (eds) Beyond the curse: Policies to harness the power of natural resources. International Monetary Fund, Washington DC, p 1–4

Mehlum H, Moene K, Torvik R (2006) Cursed by resources or institutions? World Economy 29(8):1117–1131

Monk A (2011) Sovereignty in the era of global capitalism: The rise of sovereign wealth funds and the power of finance. Environment and Planning A: Economy and Space 43(3):1813–1832

ODI (2006) Reversing the curse: Five principles for beating the 'Natural Resource Curse.' Briefing Note 7. https://cdn.odi.org/media/documents/2089.pdf

Oteh A (2017) Foreword. In: Rietvelda M, Toledano P (eds) The new frontiers of sovereign investment. Columbia UP, New York, p xi–xiv

Pistor K, Hatton K (2011) Maximizing autonomy in the shadow of great powers: The political economy of sovereign wealth funds. Columbia Journal of Transnational Law 50:1–54

Poteete A (2009) Is development path dependent or political? A reinterpretation of mineral—dependent development in Botswana. Journal of Development Studies 45(4): 544–571

Ross M (1999) The political economy of the resource curse. World Politics 51(2):297–322

Rose P (2019) The political and governance risks of sovereign wealth. Annals of Corporate Governance 4 (3):147–271. https://doi.org/10.1561/109.00000019

Ross M (2001) Does oil hinder democracy? World Politics 53(3):325–361

Ross M (2015) What have we learned about the resource curse? Annual Review of Political Science 18:239–259

Sachs J, Warner A (1995) Natural resource abundance and economic growth. National Bureau of Economic Research Working Paper 5398

Shih V (2009) Tools of survival: Sovereign wealth funds in Singapore and China. Geopolitics 14(2):328–344.

Sovereign Wealth Fund Institute (2021) Top 100 largest sovereign wealth fund rankings by total assets. https://www.swfinstitute.org/fund-rankings/sovereign-wealth-fund

SWFI (2016) Primer: What is a sovereign wealth fund? https://www.swfinstitute.org/news/47645/primer-what-is-a-sovereign-wealth-fund

Truman E (2010) Sovereign wealth funds: Threat or salvation? Peterson Institute for International Economics, Washington, DC

van der Ploeg F (2008) Challenges and opportunities for resource-rich economies. Oxford Centre for the Analysis of Resource Rich Economies, Working Paper No. 005

Yamada M, Herto S (2020) Introduction: Revisiting rentierism—With a short note by Giacomo Luciani. British Journal of Middle Eastern Studies 47(1):1–5. https://doi.org/10.1080/13530194.2020.1714267

Yi-chong X (2010) The Political Economy of Sovereign Wealth Funds. In: Yi-chong X, Bahgat G (eds.) The Political Economy of Sovereign Wealth Funds. Palgrave Macmillan, London, p. 1–25

The Alaska Permanent Fund and the Alberta Heritage Savings Trust Fund: Divergent Paths, Divergent Outcomes

Peter J. Smith

Introduction

In the 1970s, the Organization of the Petroleum Exporting Countries (OPEC) oil crisis resulted in a spike in oil prices, creating surplus revenues in oil-producing states, including sub-jurisdictions of federal states such as Alaska and Alberta. This led to the creation of each state's respective natural resource funds (NRFs), the Alaska Permanent Fund (APF), and the Alberta Heritage Savings Trust Fund (AHSTF), in 1976. Scholarship on these funds has tended to focus on three main areas of inquiry: fund objectives, governance, and investment practices. Each respective fund had different objectives, which are partial reflections of important differences in constitutional structure. Alberta, for instance, has a parliamentary system of government, which concentrates power in the executive (cabinet), while Alaska has a constitutional system based on the, with a fragmented and divided system of government in which checks and balances and separation of powers are subordinate to a written constitution.

P. J. Smith (✉)
Athabasca University, Athabasca, AB, Canada
e-mail: jays@athabascau.ca

E. Okpanachi and R. Chowdhari Tremblay (eds.), *The Political Economy of Natural Resource Funds*, International Political Economy Series,
https://doi.org/10.1007/978-3-030-78251-1_2

Moreover, although both Alaska and Alberta are part of federal systems, these systems differ. Canada's federal system, for example, is much more decentralized than the USA, with provinces possessing greater autonomy than states. The existence of the U.S. Senate provides a voice for states and regions at the national level, which the Canadian system does not possess (Pretes 1988). Finally, significant differences in terms of each state's historical experiences and respective political cultures can help explain their respective differences in terms of fund objectives, governance, management, and performance.

Pre-Establishment Phase—The Alberta Heritage Savings Trust Fund (AHSTF)

The phenomenon of Western alienation is critical to understanding Alberta's political economy and political culture, both of which can be said to have influenced the AHSTF. According to Elton and Gibbins, "Western alienation is a regional political ideology of discontent" (1979, p. 83) whose roots lie in the quasi-colonial status of Alberta and Saskatchewan when they entered Canadian Confederation as provinces in 1905. At the time, the central government in Ottawa insisted on the retention of public lands and resources in Alberta and Saskatchewan, but even when they were granted control of their public lands and resources in 1930, the feeling of alienation did not subside. Alberta continued to perceive a lack of control and influence on national politics, as well as a sense of being merely a resource colony and market for goods produced in Central Canada, as the "Milch Cow" image in Fig. 1 exemplifies (Henry 2000). The ability to develop economically was also restricted by the control of Central Canadian banks over finance and capital, which complicated access to capital for development (Richards and Pratt 1979).

The creation of the AHSTF in 1976 by the Conservative government, led by Peter Lougheed, can thus be viewed as a symbol of Alberta's independence from the economic and political domination of Central Canada. Prior to the ascension of the Conservatives to power, Alberta had been governed by the Social Credit party, whose political base was rooted in the rural population. The Social Credit government (1935–1971) was content to be a passive collector of rents from oil production. This approach was rejected by the Lougheed Conservatives, whose power base in the newly urbanized Alberta had flourished and grown since the discovery of oil in 1947. In this regard, Richards and Pratt (1979)

Fig. 1 Milch Cow (*Source* Glenbow Museum, https://www.glenbow.org/exhibitions/online/libhtm/milch.htm. Accessed 27 Oct 2019)

speak of an "arriviste bourgeoisie" composed of a new middle class that included a business class of owners and managers of large Alberta corporations, a class of professionals such as lawyers, engineers, and scientists, and a state-administrative elite.

This new middle class rejected the approach of the Social Credit government in simply collecting rent in favor of economic diversification. Once in power, the Lougheed government raised royalty rates in 1972 to capture more rent from oil. The government also made it clear in 1974 that its primary economic objective was to move the province from "a primary extractive economy, where our resources are exported to other parts of Canada and the rest of the world, to an industrialized economy" (Lougheed, quoted in Morton and McDonald 2014, p. 3).

In principle, the creation of the AHSTF would, in part, assist in the diversification and development of Alberta's economy. With the enormous oil revenues that flowed into the Alberta treasury, there would also be enough to save for a rainy day when oil and gas revenues would inevitably run out. In April 1976, after an overwhelming election victory the previous fall in which the Progressive Conservative Party won 62% of the votes and 69 of 75 seats, the legislature passed the Alberta Heritage Savings Trust Fund Act. The idea of the fund per se was not contested, but important aspects of the legislation were. For example, the legislation stipulated that 30% of Alberta's non-renewable resource revenues be transferred to the fund but was silent on future years, to which Grant Notley, leader of the New Democratic Party and its sole member in the house, asked, "is it the position of the government that after the initial period of time has passed, there will in fact be a request each year for legislative approval for money to be transferred to the heritage savings trust fund?" (1976). Premier Peter Lougheed answered that annual authorization of the transfer to the Alberta treasury would ensure that the Legislature "controls the tap" (Alberta Legislative Assembly 1976, p. 829). This, in fact, meant that revenues transferred could vary, and that the 30% was not fixed; indeed, there was no hard, fixed contribution rate, an issue that later significantly affected the fund's growth and performance. The act also specified the objectives, organization, management, and investment policies of the Heritage Fund, capitalizing it with an initial amount of $1.5 billion.

Perhaps the most contentious aspect surrounding the AHSTF was the government's plan to have the cabinet administer the fund's assets. Notley noted that "it is undemocratic for the Cabinet ... to have authority over 80 per cent of the fund." He presciently stated "there is a real danger that it will become a slush fund for the Government" (1976). Notley's concerns were echoed by Edmonton City Council, who wanted the cabinet investment committee to be expanded to include representation from municipalities. One Edmonton alderman, David Leadbetter, stated that "the potential for corruption is very great" (1976).

The establishment of the Alberta fund by ordinary legislative action is, of course, consistent with Canada's parliamentary system, as is cabinet dominance, particularly in Alberta with its history of one-party dominance and a weak opposition. The cabinet itself at the time was dominated by Peter Lougheed, who governed in the style of previous strong charismatic Alberta premiers. As Alvin Finkel (2012) notes, Lougheed brooked no opposition to his leadership. While in principle the Legislature would "control the tap," the Alberta cabinet was given the authority to invest and manage most of the fund's assets without prior approval of the Legislature. This left the fund's direction in the hands of politicians, particularly Lougheed himself, exposing it to the vicissitudes of political life in a way that the Alaska fund never was.

Pre-Establishment phase—The Alaska Permanent Fund

Although politicians in Alaska talk frequently of economic diversification, the prospect of its occurrence has historically been minimal. As Brown and Thomas note, "given Alaska's geographic location, climate and high cost of labor, diversification will be unsuccessful unless supported by massive expenditures" and any attempt to develop Alaska's "natural resources is just a stop-gap measure at best" (1994, p. 40). Moreover, Alaska's population at the time of the Prudhoe Bay discovery was small, approximately 300,000 in 1970 (US Department of Commerce), less than ten percent of Alberta's at that time. Since local, state, and federal governments have historically been the largest employer in Alaska (Drutman 2015), including a large number of military employees, most of whom are transient, the discovery did not lead to the formation of a new middle class as had happened in Alberta. In addition, Alaska's highly individualistic political culture was market-oriented and not disposed to state-led economic development

Alaska, in effect, had few options to pursue that would mitigate its boom/bust economy to be led by oil. One option that became accepted was to put Alaska's natural resource revenues into what became known as a permanent fund. The idea of a permanent fund dates back to the early 1970s when the State of Alaska received and quickly spent $900 million from the sale of Prudhoe Bay leases. According to Anderson, "Alaskans were aghast that they had frittered away so much in so short a time. Fears of uncontrolled legislative spending had been confirmed, and Alaskans sought ways to protect their natural resource revenues for future generations" (2002, pp. 58–59). This left Alaskans with a stark lesson on how swiftly large sums of money could disappear if state spending habits were not restrained by some mechanism, an excellent example of social or policy learning. According to Hall, "learning is conventionally said to occur when individuals assimilate new information, including that based on past experience, and apply it to their subsequent actions" (1993, p. 278). Others, such as Hugh Heclo, have argued that policy learning has been most frequently driven by "the perceived failings of past policy" (1974, p. 303).

The result of this lesson was a broad social and political consensus (Brown and Thomas 1994, p. 38) that part of the forthcoming revenues from the giant Prudhoe Bay field be placed in a permanent fund, an idea championed by Republican Governor Jay Hammond, who is now considered the father of the fund. In November 1976, by a two-to-one margin, Alaskans voted for an amendment to their state constitution which specified that "at least twenty-five per cent of all mineral lease rentals, royalties, royalty sale proceeds, federal mineral revenue-sharing payments and bonuses received by the state shall be placed in a permanent fund" (Alaska Permanent Fund Corporation 2021). The amendment excluded revenues from severance taxes, meaning, in effect, that the Permanent Fund would receive only 10% of the state's petroleum revenues. The amendment also left it to the legislature to determine the objectives of the fund and the disposition of the fund's earnings. The particular method by which the Alaska Permanent Fund was established was dictated by the state's constitution, which prohibited a dedicated fund, thus requiring a constitutional amendment to create the Permanent Fund. The amendment gave the fund a degree of protection it would not have had in a parliamentary system.

Establishment Stage: The AHSTF and the APF

The choice of fund objectives was as important as the decision to establish the funds. In both cases, potential objectives of the funds discussed at the time were economic, social, and fiscal/savings. Economic objectives were intended, for example, to promote economic diversification and development by providing infrastructure, subsidized loans, or high-risk equity investments. Social objectives included initiatives to change social conditions ranging from wealth redistribution to meeting other social and public facility needs. The third objective intended the funds to be used as a savings account, to maximize return, or to reduce government debt. Those advocating the savings account concept saw the funds as a future income stream to offset the near-certain decline of petroleum revenues.

In Alberta, it is evident that from the beginning, the government thought the fund could somehow pursue all three objectives simultaneously. Introducing the AHSTF bill to the legislature on April 23, 1976, Premier Lougheed outlined four priorities of the proposed fund: to serve as a savings account for a future source of revenue to offset resource revenue declines; to lessen the need for the province to borrow and thus reduce the province's future debt load; to improve the quality of life; and to strengthen and diversify the economy. Lougheed stressed, however, that both the economic and savings objectives were of equal importance and would not be easily met (Alberta Legislative Assembly 1976, p. 829). The initial emphasis on economic diversification is reflective of Alberta's historical desire to break out of the cycle of economic booms and busts common to the prairie provinces, as well as to make Alberta more economically and politically independent from Central Canada. With a strong energy sector and a strong agricultural sector to build upon, the government hoped that the economy could expand into petrochemicals, the growing information and medical research industries, tourism, forestry, and coal production. In 1980, the government, less enchanted by the prospects of economic diversification, amended the fund's objectives, no longer requiring it to strengthen *and* diversify the economy, but only to strengthen *or* diversify the economy.

Internally, these objectives were in tension with one another. Though Lougheed put heavy emphasis on the savings account, in reality the economic objectives of the fund dominated. 65% of the fund was invested between 1976 and 1982 into the Alberta Investment Division (AID) of the fund, whose mission was to strengthen or diversify the Alberta economy (Morton and McDonald 2015). Some, such as Bellingham (2017), question the wisdom of requiring the fund to be both a savings account and a means of diversification: "You do not protect savings by investing in the very economy that is vulnerable to the shocks from which you are looking to protect future generations" (p. 57).

In a variety of respects, the authorization of these objectives permitted the cabinet, functioning as the Investment Committee of the fund, to elude legislative scrutiny of government spending. For example, the third objective of improving the quality of life included everything from parks to housing loans, health care, library books, and scholarships, all of which would have been provided for by the traditional budgetary process in which the government is accountable to the legislature.

In Alaska, the process of selecting the objective of the Permanent Fund took over three years, involving widespread public debate. This process was consistent with Alaska's political culture, which, according to McBeath (1987), displays "a greater sense of political participation" (p. 78). Public participation, however, was just one reason for the delay in selecting the fund's objective; the major reason was the difficulty in reaching a consensus in the state legislature. The fragmented nature of the American political system, its separation of powers, and its weak party system make resolving political differences difficult. This was particularly true in Alaska during the late 1970s.

Each of the objectives outlined earlier had its proponents both inside and outside of the legislature (Morehouse 1984b). Initially, the economic development objective appeared to have the most support. Support for the social welfare objective, on the other hand, was limited principally to Governor Jay Hammond, who argued that earnings from the fund should be distributed as dividends. Hammond felt that dividends would ensure not only that all Alaskans would be beneficiaries of the state's wealth, but also that there would be a constituency created with an interest in protecting the fund (Alaska 1981). However, the key struggle was not over earnings, but over investment of the fund's principal between the proponents of the economic and fiscal/savings objectives.

The proponents of economic development were primarily based in the State Senate. while the fiscal/savings objective found its legislative champion in the State House of Representatives. The latter took the lead in shaping public opinion, particularly through its adept use of consultants' reports. Sound economic growth, the consultants argued, was driven by the market, not public policy, a view close to the market ideology of most Alaskans (Morehouse 1984b, pp. 172–173). The consultants also argued that the fund should be operated as a trust and invested outside Alaska, with managers accountable to the public through the legislature and governor and public reporting requirements.

Ultimately, this was the view of the fund that prevailed. The economic objective, the most contested, was eventually abandoned as legislation stipulating the savings objective as appropriate for the Permanent Fund passed both houses of the legislature in April 1980. The legislation provided for investment management and accountability, but was silent on the disposition of the fund's earnings, a question that was not resolved until 1982.

Organization, Governance, and Investment of the AHSTF and the APF

The contrast in objectives is representative of other significant differences between these two funds. Unlike Alaska's Permanent Fund, which is managed and invested as a single unit independent of government interference, the governance and management of Alberta's Heritage Fund was vested in the Cabinet, who made the major investment decisions and set the fund's direction. Structurally, the Heritage Fund was originally divided into five investment divisions of varying size and purpose: the Alberta Investment Division (AID); the Capital Projects Division; the Canada Investment Division; the Commercial Investment Division; and the Energy Investment Division. The first three were created by the original legislation in 1976, and the latter two were added in 1980. At the time, no long-term goal of an annual real rate of return for the fund had been publicly announced. On a day-to-day basis, the Treasury Board was responsible for managing most of the fund's assets in its various divisions. In certain special instances, departments were given responsibility for overseeing equity investments in certain areas. For example, the Department of Energy and Natural Resources was given responsibility for monitoring AID's investment in the Syncrude oil sands in northern Alberta (Pritchard 1976).

By far, the most significant division of the fund was the AID. Investments in AID, as determined by the Cabinet, were required to strengthen or diversify the economy of Alberta and yield a reasonable return or profit. No limitation was placed on the size of the division. During the economic slowdown of the early 1980s, the AID formed part of an interventionist strategy of Lougheed's successor, Conservative Premier Don Getty, who made it clear that he favored government intervention, declaring "Government is a large stimulator of the economy... I am not going to wait for the banks" (quoted in Morton and McDonald 2015, p. 15). Many, but not all, of the investments, including loans, equity, and loan guarantees, were unsuccessful and politically unpopular, leading to a reformulation of the fund's structure and investment policies in 1996 under Getty's Conservative successor, Ralph Klein.

The Canada Investment Division (CID) consisted of loans to provincial governments and crown corporations. Underlying the division was a political motive. According to Pretes, "the growth of the Alberta energy industry, manifest in the rapid growth of the Alberta Heritage Fund, ... brought the province into open conflict with the federal government" (1988, p. 45). Ottawa, in turn, was being pressured by the province of Ontario, whose taxpayers were providing 29% of fund revenues (Pretes 1988); as Simeon stated, Ontarians saw "blue-eyed Arabs bleeding them dry" (1980, p. 184). The purpose of the CID, then, was to assuage Canadian public opinion and that of the federal government, which viewed Alberta's wealth as a potential threat to the distribution of power within the Canadian political system (Pretes 1988). The Commercial Investment

Division focused on equity investments expected to yield a commercial return or profit.

The Energy Investment Division was established in 1980 to facilitate the development of energy resources within Canada, but it never had any investments of consequence. Next to the AID, the most controversial division of the fund was the Capital Projects Division (CPD), which spent billions for such projects as parks, hospitals, oil sands research, and irrigation ditches, all of which would ordinarily have been allocated through the traditional budgetary process. This division, which could include up to 25% of the fund's assets, was intended to provide long-term economic or social benefits and was not expected to yield a return.

Contributions to the fund have altered significantly over the years. Prior to 1983, the Heritage Fund received 30% of all resource revenues, but the recession of 1982 prompted the government to reduce that figure to 15%. In fact, yearly transfers to the fund ceased in 1987 and did not resume until 2007, with all earnings of the fund going into the general revenue account. In addition, nearly $40 million of investment income have been transferred to the General Revenue Fund. Capital project expenditures ceased as of 1996 (AHSTF Annual Report 2019).

Alaska's fund experienced significant differences from Alberta's, in terms of not only objectives but also management and organization. Alaskan legislators, for example, felt that the Permanent Fund should not be left under the direct control of the legislature or executive, but still had to be accountable to elected officials. According to a legislative report, "the aim was insulation without isolation" best achieved by creating "a public corporation distinct from state government" (Alaska 1980).

Within the management framework outlined by the legislature, corporate policy is formulated independently by a board of six trustees, appointed by the governor at his/her discretion for a four-year term, four of whom are members of the public with recognized competence in business and finance (Alaska 1980). The others are cabinet officers, one of whom must be the Commissioner of Revenue. The board employs an executive director who is responsible, subject to board approval, for hiring additional staff. Together, the executive director and staff carry out the day-to-day operations of the fund. The board also has access to an advisory group composed of former trustees and financial experts to assist them in evaluating the advice of staff and consultants.

Structurally, Permanent Fund equity is composed of two parts: principal and earnings reserves. The principal of the Permanent Fund as of

July 31, 2019, stood at $47.9 billion (Alaska Permanent Fund Corporation 2019). The principal of the fund includes oil royalties deposited by constitutional dedication; extra money deposited by legislative appropriation; and income of the fund that has been transferred to principal in order to "inflation-proof the fund."

Income from the fund is either distributed as yearly Permanent Fund Dividends transferred to principal for inflation-proofing, or placed in the Earnings Reserve Account (ERA). The ERA is managed as part of the fund, but can be appropriated by ordinary act of the legislature. As of July 31, 2019, the ERA stood at $18.5 billion; the total fund value was thus $66 billion (US) in 2019 (Alaska Permanent Fund 2019). Expenditure of the principal, however, is prohibited by the state's constitution. Not until 1982 was there a requirement that effects of inflation on the principal be offset by an automatic reinvestment of income each year, with the amount determined by the Consumer Price Index.

As managers of a public trust, board members must follow the "Prudent Investor Rule," meaning that it is more important to invest the money safely than to maximize investment earnings. The long-term goal of the trustees is to realize an annual earnings rate of 5% above inflation, which it has consistently met (APFC Annual Report 2019). The investment policy of the fund has broadened over time as confidence in fund management has grown to include investing globally—but not within Alaska, unless it meets the criteria of the Fund—with the majority, $47.9 billion (73%), invested in the USA (APFC Annual Report 2019). At present, the target asset allocation of the fund is primarily in public equities and fixed income, plus lesser amounts in private equity, real estate, and other forms of investment (APFC Annual Report 2019)

Natural Resource Funds in Operation—Alaska and Alberta

As critical as organization, management, and investment of these funds are, the question of accountability, the means of ensuring that elected politicians and public servants perform their functions in the public interest, is equally important. Effective accountability in these cases relies primarily upon the technical capacity of the legislature to provide oversight and upon public knowledge of the funds and their activities. Technical capacity here refers to the "financial expertise to understand and assess activity occurring in large investment functions" (Helgath and

Bibb 1986, p. 57). Although both funds suffer from a relative lack of legislative oversight, Alaska does have a public that scrutinizes the fund and holds politicians accountable for its management, where Alberta does not.

In formal terms, while a standing committee of legislators had the responsibility to provide oversight of the AHSTF, the committee was controlled by government MLAs who in turn took their cues from the provincial cabinet. From its inception, the committee has been ineffective. When committee members challenged the government, particularly on the need for a public review of the fund, the cabinet ignored them, frustrating even some government members of the committee (Alberta Legislature Standing Committee 1986, p. 339).

Weak legislative accountability could potentially be remedied by an informed and knowledgeable public. This, however, was not the case in Alberta. In his appearance before the committee in 1986, Premier Don Getty admitted that he could "find very few people, in public or private life, who express ... any real knowledge of the fund" (Alberta Legislature Standing Committee 1986, p.35). Not much has changed in the intervening years (Bellingham 2017).

The American approach to accountability, with its weak party system and fragmentation of power and openness, is distinct from the Canadian system, which concentrates power in the cabinet. In the USA, the public service is expected to answer not only to the legislature but also to the executive and the public. According to the state legislature, the Permanent Fund was intended to be "protected from political influences, but at the same time, responsive to changes in state policy and accountable to the people through their elected officials" (Alaska 1980). The Legislative Budget and Audit Committee (LBA), a bicameral body, was authorized by statute to oversee the fund. In 1981 and 1982, a subcommittee of the LBA did considerable work on the fund, particularly on principles guiding the Fund's investments, but overall there has been little legislative oversight. Each year, the LBA regularly accepts the annual audit of the Permanent Fund. The Alaska Permanent Fund Corporation (APFC) is also partially accountable to the governor, who appoints and can dismiss trustees. To date, however, no governor has exercised the option to dismiss trustees, nor has any administration developed the technical capacity to assess the fund. The deficiencies of legislative and administrative oversight are, in part, remedied by the fund's public accountability by, for example, making independently audited statements, along with a statement of earnings from each investment and a market-value appraisal of the investments, available to the public and written in easily understood language (Alaska 1980). In 1982, the legislature moved to improve accountability by requiring that part of the fund's earnings be distributed annually as a Permanent Fund Dividend (PFD). This move was intended both to ensure that Alaskans of all ages receive, as owners, a share of the oil wealth and to encourage greater public scrutiny of the fund's management.

The PFD was championed by Governor Jay Hammond, who believed that all Alaskans, as owners of the oil—but not the oil companies themselves, which are all private—should receive an annual benefit, a dividend from the fund's earnings. Hammond hoped that the dividend would limit government spending and encourage Alaskans to keep an eye on the fund as they had a vested interest in its performance (Anderson 2002). However, raids on the fund by greedy politicians became a common fear. According to David Rose, the first director of the APFC, the greatest threats to the fund are the "intense temptations of powerful people (and) ... the strongest defense of all is the Alaska Permanent Fund dividend" (quoted in Hickel 2019).

As a means of accountability, the PFD has worked, perhaps too well, as, according to Brown and Thomas (1994), it was soon regarded as an entitlement, a "sacred political cow" that no politician dare touch. Organizations such as the Permanent Fund Defenders (2019) have been determined to "Save Our PFD and Permanent Fund from Attacks on Alaskans' Common Wealth." Today, with Alaska facing large budget deficits, political battles over the size of the fund's annual dividend have occurred. Until the size of the PFD was trimmed in 2018, money for the dividend had been calculated on a formula based on the realized net income for the previous five years (Erickson and Groh 2012, p. 42). The size of the dividend has risen substantially, but since fund earnings are dependent on market forces, there have been periods of decline, with the peak years in about 2000 when they exceeded $2500 US per person (adjusted for inflation).

Beyond the issue of accountability, it is important to examine how well these funds respectively performed and lived up to their original purposes. Table 1 provides an overview of the very different directions the AHSTF and the APF took.

In a fundamental sense, the AHSTF was hampered from the beginning by the fact that the savings and economic diversification objectives were

Table 1 Some similarities and differences—The AHSTF and the APF

	Alberta	*Alaska*
Invest internally (in province/state)?	Yes	No
Individual dividend?	No	Yes
Channeled through Treasury/Dept of Revenue?	Yes	Yes
Economic objective?	Yes (pre 1997)	No
Social objective?	Yes (pre 1997)	No
Savings objective?	Yes	Yes
Unified fund?	No	Yes
Reinvest returns?	No	Yes (for inflation)
Provincial/State Public property right regimes for minerals/oil/gas?	Yes (the Crown)	Yes, but only on state-owned land (eg. Prudhoe Bay)

Source This table has been adapted from personal correspondence with Jonathan M. Moses of the Norwegian University of Science and Technology

in conflict with one another, leading to contradictory government policies regarding these objectives. The posing of the economic development objective stemmed from a strongly held desire by Albertans to use the Fund for economic development. At the same time, the unwillingness of Albertans to see the Fund invested outside the province hamstrung the government's ability to have the type of sound portfolio management so necessary to optimize the savings objective.

When the Conservative government engaged in economic diversification projects, more often than not these projects ended in failure. Finkel (2012) and Morton and McDonald (2015) have itemized these failures, with the latter creating a list of failed diversification projects during the Lougheed and Getty years. To be fair, there were some notable successes, such as providing loans to Syncrude, which has gone on to play an instrumental role in the development of the oil sands in northern Alberta. Even so, the public mostly remembered the losses and the increasing provincial debt wracked up by the Conservatives to maintain public services.

By the early 1990s, fund-led diversification had lost popular and political favor, which, coupled with the growing provincial debt, led to public demand for a new approach to governing. The result was what Hall has described as a third-order change, a paradigmatic shift "marked by the radical changes in the overarching terms of policy discourse" (1993, p. 279). The government interventionist approach of the Lougheed and Getty governments was replaced by the market-oriented, neoliberal approach of the Conservative government led by Ralph Klein (Harrison and Laxer 1995).

What followed was a reconceptualization and reform of the AHSTF beginning in 1996, including a dissolution of the fund's divisions and holdings to portfolio management for long-term gain in portfolio value. In the 1997–1998 annual report of the fund, the provincial treasurer stated: "Based on the advice of Albertans, the focus of the Fund is now on providing the greatest financial return on investments. The Heritage Fund no longer invests in specific projects with economic or social goals and existing investments will be reduced through time" (Day 1998, p. 4). In addition, the fund was expected to retain enough from earnings to remain inflation-proof. However, this was subject to the whim of the government and did not start until 2006. Responsibility for the investment of the fund was provided by a division of the Treasury Board.

In yet another show of reform in 2007, the government created the Alberta Investment Management Corporation (AIMCo), which manages not only the assets of the AHSTF but also public sector pensions and other public endowments and funds. AIMCo functions as a provincial board that operates outside the strictures of government, including the cabinet and legislature—in effect, as a privatized entity (Ascah 2016).

Despite these changes, the size of the fund has not increased. Earnings of the fund continue to be channeled to the general revenue to support higher spending. Since its inception, the fund has earned $43.5 in income with $40 billion Cdn transferred to the General Revenue Account (AHSTF Annual Report 2019). Ascah notes that savings have continued to be a problem in Alberta (2013, p. 189); however, the hope that the Fund might be reformed with Alaska as the model still exists. In their analysis, "Fumbling the Alberta Advantage," Milke and Palacios (2015, p. 2) argue that Alberta should have followed Alaska's lead:

> Had Alberta imitated the state of Alaska ... and followed the Alaska rule that requires 25% of all resource revenues to be deposited into the Alaska Permanent Fund, from 2005/06 to 2013/14 inclusive, instead of just $4.5 billion in actual deposits, the deposits would have amounted to $25.3 billion.

Elsewhere, Murphy and Clemens (2015) lament the inability of Alberta to save and invest more in the AHSTF. Drawing lessons from Alaska, they urge reforms "including a formal rule for the contribution percentage and institutional mechanisms to encourage proper fund management" (p. 5). However, the chances of Alberta creating a formal rule for contributions to the fund are remote for many of the reasons Ascah discusses. In addition, oil prices are currently middling at best, dropping from over $100 a barrel in 2014 to under $60 in 2019, which are too low for Alberta to start saving again.

The question here is whether Alberta has been able to escape the resource curse; the answer is highly debatable. Gelb (2014), for example, argues that those "who compare this country to illiberal and corrupt 'petro-states' are being either ignorant or deceitful" (p. i). On the other hand, the title of Taft's book *Oil's Deep State: How the Petroleum Industry Undermines Democracy and Stops Action on Global Warming—In Alberta, and in Ottawa* (2017) sums up his argument. Shrivastava and Stefanick (2015) agree that the petroleum industry has contributed to a democratic deficit in Alberta. Nonetheless, other factors may be at play: Alberta has long had long periods of one-party rule and weak opposition; in addition, neoliberalism can be said to have contributed to inequality with its corroding effects on democracy.

In contrast, there is little doubt that the APF has been successful in terms of its meeting its savings objective. Its primary objective of being a savings account gave it a clarity of purpose that the AHSTF never achieved, and has simultaneously facilitated public accountability and support. Public support of the Permanent Fund was facilitated by its "de-politicization"; that is, by the decision to create a public corporation distinct from the government. Management of the fund by experts in finance and business has reinforced public perception of its competence and political neutrality, permitting the trustees to invest increasingly in common stocks, which over time yield a higher rate of return. In recent years, the APF has received a number of international awards for its management and investment strategy (Annual Report 2018). In sum, the APF has fulfilled its purposes of being a savings account and providing an income stream.

The Permanent Fund Dividend has, in particular, been singled out for praise. Widerquist and Howard (2012) note that its two main benefits are that it saves money for future generations and for a day when oil is exhausted, and that it mitigates poverty and inequality. In the 2019 Democratic presidential campaign, Andrew Yang attracted considerable attention for the idea of giving every American over the age of 18 a

"Freedom Dividend" of $1,000 per month, a notion that Yang admits was indebted in part to the APF (Savchuk 2018).

Although Widerquist, Howard, and Yang have praised the PFD, the future of the dividend has led to political conflicts in Alaska. Faltering oil prices have led to a strain on the state's budget, which is constitutionally prohibited from running a deficit. In the spring 2019 legislative session, contestation over budget cuts spilled over into a debate on the PFD. Representative Bryce Edgmon lamented "[t]hat we can't continue to allow it [PFD] to dominate the political landscape here in Juneau and across Alaska" (Maguire 2019). However, Edgmon's opinion seems to be wishful thinking, as the issue of what to do with the dividend has not been settled. Some support a change in the statutory formula determining the size of the dividend, but both the political right and left support keeping the formula as is. Prominent Republicans see maintaining the dividend's size as a means of limiting the size of government (Moniak 2019), while Democrats support maintaining the formula for different reasons. Higginbotham (2019) argues that the dividend particularly benefits the poor:

> 79 percent of Alaskans call the PFD an important source of income, while the demographics it benefits the most are mothers of young children, unmarried women, women without a college degree, and Alaska Native women. A 2016 study found the dividend reduces poverty in the state by 20 percent.

Has Alaska managed to avoid the resource curse? As in the case of Alberta, opinions are mixed. Widerquist (2012) argues that Alaska has avoided both the "Dutch Disease" and corruption primarily by means of the APF and the PFD: the former primarily by investing outside the state, thus lowering the opportunities for corruption, and the latter by returning dividends to the people instead of concentrating it in the hands of a few. On the other hand, James (2016) argues that the boom in oil production in Alaska "generated significant short-run economic gains that were quickly diluted by inward migration" (p. 270). Moreover, "as oil production and prices fell ... the income gains had even turned into losses" (James 2016, p. 275). James also notes, contrary to Widerquist, that an investigation by the US Department of Justice found public corruption in Alaska, which led to the conviction of multiple past and present state representatives for accepting illegal bribes from an oil services company.

Conclusion

As other contributors to this volumehave pointed out (see Moses), there has been a tendency to view NRFs as apolitical technical means of investment administered by market experts. Yet, this approach leaves too much out. Politics, constitutional structure, history, and political culture provide vital context on how the AHSTF and APF were created, structured, and performed. Politics, for example, influenced almost everything about the AHSTF, perhaps to its detriment, resulting in a fund that never lived up to its promise. The APF was able to insulate itself to a certain extent from politics by creating the APFC. However, politics is constantly present and is notably visible in the present-day contest over the PFD. In addition, constitutional structures, whether presidential, parliamentary, or federal left their marks on each fund, as did history and political culture.

Today, both Alaska and Alberta face daunting futures, limiting their ability to rely on natural resource funds to augment the APF, or fund current public services from or save money in the AHSTF. The Paris Agreement of 2015 is one indicator that the fossil fuel industry may be a sunset industry; another is disinvestment in the industry, as "Trillions of dollars of investments are being taken out of carbon-intensive companies" (McKibben 2018). In Alaska, the revenue-producing Prudhoe Bay field is in decline with no replacement in sight. Although Alberta has abundant reserves in its northern oil sands, the inability to get its product to market due to increased resistance to building more pipelines to tidewater has led to a rekindling of Western alienation. Climate change is another challenge, with massive forest fires occurring in both Alaska and Alberta. In Alberta, the estimated cost of cleaning up all of the old and unproductive oil and gas wells is between $40 billion and $70 billion (CBC News 2019), which is beyond the province's current capacity.

When all is said and done, the APF will likely survive these challenges, at least in the short term, as an intergenerational trust with the Earnings Reserve Fund and the dividend being contested. Without a replacement staple, the question is how long this will last as economic and fiscal challenges mount. In Alberta, the prospect is more doubtful: the heyday of the AHSTF and large oil surpluses has passed, with little evidence that it will come again.

References

Alaska (1980) Free Conference Committee, Final Report, Apr

Alaska (1981) Interview with Jim Rhodes, special assistant to the Alaskan Permanent Fund Trustees, 29 Apr

Alaska (2019) Department of Labor and Workplace Development, Alaska population estimates. http://live.laborstats.alaska.gov/pop/. Accessed 24 Aug 2019

Alaska Oil and Gas Association (2019) More than 1/3 of Alaska's jobs are tied to the oil and gas industry. https://www.aoga.org/facts-and-figures/state-revenue. Accessed 03 Sept 2019

Alaska Permanent Fund Corporation (2019) Our performance. https://apfc.org/our-performance/. Accessed 05 Sept 2019.

Alberta (various dates) Alberta Heritage Savings Trust Fund annual report

Alberta (1976) Legislative Assembly Debates

Alberta Legislature (various dates) Standing Committee of the Alberta Heritage Savings Trust Fund Act

Anderson J (2002) The Alaska Permanent Fund: Politics and trust. Public Budgeting and Finance 22:57–68.

Ascah R (2013) Savings of non-renewable resource revenue: Why is it so difficult? A survey of leaders' opinions. In: Ryan D (ed) Boom and bust again: Policy Challenges for a commodity-based economy. U of Alberta P, Edmonton, p 151–197

Ascah R (2016) Alberta Investment Management Corporation: An examination of corporatization project. University of Alberta Department of Economics, Institute for Public Economics

Alaska Permanent Fund Corporation (2021) History of the Alaska Permanent Fund. https://apfc.org/who-we-are/history-of-the-alaska-permanent-fund/

Bellingham D (2017) The electoral temptations of natural resource revenues: Towards a theory on the origins of sovereign wealth funds in Canada. MA thesis, Carleton University

Bohrer B (2019) Embattled Alaska governor scales back budget cuts, approves oil-wealth dividend of $1600. NBC News. https://www.nbcnews.com/news/us-news/embattled-alaska-governor-scales-back-budget-cuts-approves-oil-wealth-n1044446. Accessed 31 Aug 2019

Brooks J (2019) Dividend dispute presents and elephant-sized dilemma for Alaska lawmakers in special session. Anchorage Daily News. https://www.adn.com/politics/alaska-legislature/2019/05/18/dividend-dispute-presents-an-elephant-sized-dilemma-for-alaska-lawmakers-in-special-session/. Accessed 26 May 2019

Brown W, Thomas C (1994) The Alaska Permanent Fund: Good sense or political expediency? *Challenge*. September/October:38–44.

CBC News (2019) Old, unproductive oil and gas wells could cost up to $70B to clean up, says new report. https://www.cbc.ca/news/business/orphan-wells-alberta-aldp-aer-1.5089254. Accessed 08 Dec 2019

Collins A (1980) The AHSTF: An overview of the issues. Canadian Public Policy 6: Supplement, 158–166.

Day S (1998) Alberta: Annual report of the Alberta Heritage Savings Trust Fund.

Di Loberto T (2019) High temperatures smash all-time records in Alaska in early July 2019. Climate.gov. https://www.climate.gov/news-features/event-tracker/high-temperatures-smash-all-time-records-alaska-early-july-2019. Accessed 28 Aug 2019

Dittman P (2019) Alaska public policy issues. https://www.alaskapublic.org/wp-content/uploads/2019/08/Alaska-Public-Policy-Opinion-Survey-Results-May-2019.pdf. Accessed 25 Aug 2019

Drutman L (2015) Alaska: Land of contradictions. Pacific Standard. https://psmag.com/politics-and-law/alaska-land-of-contradictions-4209

Elton D, Gibbins R (1979) Western alienation and political culture. In: Schultz R, Kruhlak O, Terry J (eds) The Canadian Political Process, Third Edition. Holt, Rinehart, and Winston, Toronto, p 82–97

Erickson G, Groh, C (2012) How the APF and the PFD operate: The peculiar mechanics of Alaska's state finances. In: Widerquist K, Howard M (eds) Alaska's Permanent Fund. Palgrave Macmillan, New York, p 41–49

Finkel, A (2012) Myths communicated by two Alberta dynasties. In: Taras D, Waddell C (eds) How Canadians communicate IV: Media and politics. Athabasca UP, Edmonton, p 189–211

Gelb A (2014) Should Canada worry about a resource curse? University of Calgary School of Public Policy Research Paper 7(2):1–26

Hall P (1993) Policy paradigms, social learning and the state. Comparative Politics 25(3):275–296

Harrison T (2015) Petroleum, politics, and the limits of left progressivism in Alberta. In: M. Shrivastava M, Stefanick L (eds) Alberta oil and the decline of democracy in Canada. Athabasca UP, Edmonton, p 70–89

Harrison T, Laxer G (1995) Introduction. In: Laxer G, Harrison T (eds) The Trojan Horse. Black Rose, Montreal, p 1–22

Helgath S, Bibb S (1986) Alaska's Permanent Fund: Legislative history, intent and operations. Alaska State Senate, Juneau

Heclo H (1974) Modern social politics in Britain and Sweden. Yale UP, New Haven

Henry S (2000) Revisiting western alienation: Towards a better understanding of political alienation and political behaviour in western Canada. Dissertation, University of Calgary

Hickel J (2019) Defend the Permanent Fund and dividend, or lose them both. Anchorage Daily News, 13 Apr. https://www.adn.com/opinions/2019/04/

13/defend-the-permanent-fund-and dividend-or-lose-them-both/. Accessed 05 May 2019

Higginbotham T (2019) The wrong way to end austerity in Alaska. Jacobin Magazine, 19 July. https://www.jacobinmag.com/2019/07/alaska-permanent-fund-dividend-dunleavy. Accessed 24 Aug 2019

Innis H (1984) The importance of staple products. In: Easterbrook W, Watkins M (eds) Approaches to Canadian economic history. Carleton UP, Ottawa, p 16–19

James A (2016) The long-run vanity of Prudhoe Bay. Resources Policy 50(C):270–275

Johnson S (1987) The Alaska legislature. In: McBeath G, Morehouse T (eds) Alaska: State Government and Politics. U of Alaska P, Fairbanks, p 135–166

Leadbetter D (1976) Globe and Mail, 24 Mar, p 31

Maguire S (2019) Dozens call opposing House PDF plan that would change the dividend formula. KTUU. https://www.ktuu.com/content/news/Dozens-call-opposing-House-PFD-plan-that-would-change-the-dividend-formula-510359851.html. Accessed 26 May 2019

McBeath G (1987) Alaska's political culture. In: McBeath G, Morehouse T (eds) Alaska: State government and politics. U of Alaska P, Fairbanks, p 7–29

McKibben B (2018) At last, divestment is hitting the fossil fuel industry where it hurts. The Guardian, 16 Dec. https://www.theguardian.com/commentisfree/2018/dec/16/divestment-fossil-fuel-industry-trillions-dollars-investments-carbon. Accessed 8 Dec 2019

McMillan M, Warrack A (1995) One track (thinking) towards deficit reduction In: Laxer G, Harrison T (eds) The Trojan Horse. Black Rose, Montreal, p 134–163

Milke M, Palacios M (2015) Fumbling the Alberta advantage. Fraser Institute, Edmonton

Moniak R (2019) Opinion: Dunleavy is still the odd man out. Juneau Empire, 23 Aug. https://www.juneauempire.com/opinion/opinion-dunleavy-is-still-the-odd-man-out/. Accessed 24 Aug 2019

Morehouse T (1984a) Introduction. In: Morehouse T (ed) Alaskan resource development. Westview, Boulder, p 1–22

Morehouse T (1984b) Resource endowments, economic markets, and public policies. In: Morehouse T (ed) Alaskan resource development. Westview, Boulder, p 209–239

Morton T, McDonald M (2015) The siren song of economic diversification: Alberta's legacy of loss. University of Calgary School of Public Policy Research Paper 8(2):1–29

Murphy R, Clemens J (2015) Reforming Alberta's heritage fund: Lessons from Alaska and Norway. Fraser Institute, Edmonton

Notley G (1976a) Globe and Mail. 24 Mar, p. 31

Notley G (1976b) Alberta Hansard, 14 Apr
Permanent Fund Defenders (2019) Save our PFD and Permanent Fund from attacks on Alaskans' common wealth. http://www.pfdak.com/. Accessed 07 Sept 2019
Pretes M (1988) Conflict and cooperation: The Alaska Permanent Fund, The Alberta Heritage Fund and Federalism. American Review of Canadian Studies 18(1):39–49
Pritchard T (1976) Cabinet to decide major Heritage Fund investments. Globe and Mail, 30 June, p B1
Richards J, Pratt L (1979) Prairie Capitalism. McClelland and Stewart, Toronto
Shrivastava M, Stefanick L (2015) Framing the debate on democracy and governance in an oil-exporting economy. In: Shrivastava M, Stefanick L (eds) Alberta oil and the decline of democracy in Canada. Athabasca UP, Edmonton, p 3–28
Simeon R (1980) Natural resource revenues and Canadian federalism: A survey of the issues. Canadian Public Policy, Supplement, p 182–191
Sran, G. (2017). Historical overview Government of Alberta revenues and expenditures. AUPE PEF Presentation Canadian Economics Association Conference, 2 June. http://calgaryhomeless.com/content/uploads/Sran-PEF_CEA_June2-2017.pdf. Accessed 03 Sept 2019
Savchuk K (2018) Frozen assets. Mother Jones. https://www.motherjones.com/politics/2018/12/alaska-oil-permanent-fund-dividend-universal-basic-income/. Accessed 07 Sept 2019
Taft K (2017) Oil's deep state: How the petroleum industry undermines democracy and stops action on global warming – In Alberta, and in Ottawa. James Lorimer, Toronto
US Department of Commerce (1973) 1970 census of population, Alaska. http://live.laborstats.alaska.gov/cen/histpdfs/1970char.pdf. Accessed 04 Sept 2019
Widerquist K (2012) How Alaska can avoid the third stage of the resource curse. Basic Income News. https://basicincome.org/news/2018/05/alaska-can-avoid-third-stage-resource-curse-2012/. Accessed 23 Oct 2019
Widerquist K, Howard M (2012) Introduction: Success in Alaska. In: Widerquist K, Howard M (eds) Alaska's Permanent Fund. Palgrave Macmillan, New York, p 3–15

The Entanglements of Natural Resource Funds: Fundo Soberano de Angola and Domestic Political Interests

Terhemba N. Ambe-Uva and Sarah J. Martin

Introduction

In October 2012, *Fundo Soberano de Angola* (the Angola Fund, or Fund) was tasked with utilizing revenues from oil "to promote growth, prosperity, and social and economic development" (IMF, 2019, p. 43) and diversify the economy (FSDEA, 2013). The Angola Fund is currently the second-largest SWF in sub-Saharan Africa (Sovereign Wealth Fund Institute, 2021). The primary aim of the Sovereign Wealth Fund (SWF) is to promote Angola's social and economic improvements and enhance equitable intergenerational transfer (Chen, 2019). The Fund judiciously implements the influential soft law standards of sovereign wealth fund behavioral principles (Backer, 2015) known as Santiago Principles—"a framework of generally accepted principles and practices that properly reflect appropriate governance and accountability arrangement as well as the conduct of investment practices by SWFs on a prudent and sound

T. N. Ambe-Uva (✉)
University of Ottawa, Ottawa, ON, Canada
e-mail: tambe093@uottawa.ca

S. J. Martin
Memorial University of Newfoundland, St. John's, NL, Canada

E. Okpanachi and R. Chowdhari Tremblay (eds.), *The Political Economy of Natural Resource Funds*, International Political Economy Series, https://doi.org/10.1007/978-3-030-78251-1_3

basis" (IWG, 2008, p. 4). The Fund was launched in the context of a rapid increase in SWFs, especially in the wake of the 2007–2008 global financial crisis as many resource-rich developing countries were confronted with a sharp drop in revenues from commodities.

The Fund's governance is designed around international best practice, along with internal and external audits and transparent investment reports. But critics assert that the Fund's loosely crafted accountability mechanisms oils the wheels of existing patronage systems (Markowitz, 2020). In turn, the Fund has become embedded in patronage networks, cronyism, and the disbursal of resources for short-term benefits without recourse to the long-term development impact, thereby derailing its key aims to foster social and economic prosperity for Angolans (Amundsen, 2014).

The new Government of President João Lourenço, elected on August 25, 2017, has embarked on a broad swath of legal reforms that could strengthen the Angola Fund's governance framework. Yet, to attribute Lourenço's policies as transformative ignores the Fund's ongoing politics, especially how the fund is embedded in President's office. This is often to neglect more in-depth explanations about how and why the President has exclusive oversight over the Fund, and how the President deploys the Fund to achieve their political agendas. To understand the Fund and its governance structure, we examine it in the context of political economic trends, and how the Fund is embedded in domestic politics, especially the office of the president.

In this chapter, we ask: how have political interests interacted with the Fund's mandates and operation? Second, what are the implications of these political interests in relation to the Fund? We argue that the Angola Fund's mandate remains highly politicized, albeit less than in the previous dispensation. We highlight how the President can bypass the Parliament to direct the Fund, and how a recent large withdrawal from the Fund has had implications for scrutiny from Parliament, and the overall transparency and accountability agenda. In addition, we suggest that the Angola Government could reinforce the Social Charter with the Fund and help finance Sustainable Development Goals in Angola and within the region. We illustrate these points by tracing the Angola Fund's historical evolution by locating it in the country's commodity-dependent status, weak institutions, the implications for the 27-year war, and the recent transition of power to President João Lourenço.

In the first section, we provide an overview of how economic fragility has shaped the Angola Fund and problematize it as the character of Angola's commodity-based economy. In the second section, we discuss the legal framework and the institutional and investment frameworks underpinning the Fund, especially the implications for recent attempts to strengthen it. We then focus on the Angola Fund's political economy as we explore how a sovereign wealth fund cannot be disentangled from the owner's politics and institutions. In doing so, we reveal that despite the initial hope that President Lourenço could reorient the Fund to serve the public interest, there are strong concerns that the Fund may not enjoy any form of independence from the presidency. Lastly, we conclude with how the Fund's domestic and regional mandates have not served the citizens' interests or diversified the economy, but as an avenue to invest in areas where the President has interests. We call for the strengthening of the Social Charter to innovatively finance Sustainable Development Goals.

Early Post-War Angola: How Economic Fragility Shaped the Angola Fund

The Fund's history cannot be divorced from the nature and character of Angola's commodity-based economy, which has remained susceptible to the vagaries of global oil and diamond prices and exogenous shocks. The rich natural resource endowments that underpin the Fund contributed to the 27-year war, which had significant impacts on human lives, infrastructures, and institutions and only ended in 2002. This section will focus on the political and economic contexts informing the Angola Fund's establishment.

The Angola Fund is closely tied to the political economy of the country. The country is in the southern part of Africa, with 29.25 million people, a GDP per capita of $3390, and a GDP at the market price of $99.15 billion (OPEC, 2019). The country has proven crude oil and natural gas reserves of 81.60 billion barrels and 383 billion cubic meters, respectively, trailing only behind Nigeria in sub-Saharan Africa (OPEC, 2019). Angola's economy is the third largest in sub-Saharan Africa, only behind South Africa and Nigeria. It is also the second-largest oil producer and the third leading producer of diamonds in the sub-region (Business Monitor International, 2015).

Angola's economy remains highly dependent on hydrocarbons and susceptible to the vagaries of external shocks. In 2018, oil and gas accounted for almost 90% of total exports (OPEC, 2019). However, the sector contributes less than 1% of the country's employment (CDP, 2018, p. 2), and a hydrocarbon-based economy is no guarantee for political and economic development, as the resource curse arguments demonstrate (Ross, 2015). Angola's governance structure, woven around hydrocarbons, has always served the interests of the privileged class. Independence from Portugal in 1975 did not alter this structure but rather deepened the resource curse (Le Billion, 2001). The 27-year resource-fueled civil war intensified and reshaped the race to extract natural resources to fund the war. Post-war Angola has extractive institutions, high poverty, and a weak democracy—despite impressive economic growth—which puts the country firmly among those cursed by resources (Amundsen, 2014). In turn, more than 70% of the population lives on less than $2 a day (Wiig & Kolstad, 2013, p. 147), and it has one of the highest infant and maternal mortality rates in the world (Jensen, 2018). Investing in new infrastructure, including health and social programs, has the potential to diversify the economy, improve the welfare of its people, and to reduce its dependence on a single commodity. These potentials have not been achieved. Instead, reliance on hydrocarbons led to economic shocks when oil prices plummeted in 2008 and then again in 2014 (Ufimtseva, 2019).

Angola's post-war economy is underpinned mainly by its political environment. The country is a multiparty democracy, and on paper, power is distributed among the executive, legislature, and the judiciary. In reality, so much power is concentrated in the presidency, where the President is the head of state, head of government, commander in chief of the armed forces, and the head of the ruling party (Vines & Weimer, 2011). Although there is parliamentary opposition, it has little or no impact on the government's composition, and the President has the exclusive rights to dissolve the Parliament and call for an election. Vines and Weimer (2011) argue that the 2010 Constitution granted the President the powers to appoint the Cabinet, which is nominally accountable to an elected National Assembly. The 2010 Constitution also changed the processes of electing a President by declaring any person named at the top of the winning party as an automatic president. Parliamentary and presidential elections happen concurrently.

The early post-war years—filled with optimism and uncertainty—saw the Angolan Government adopt a cautious approach in managing the country's natural resource revenues, by limiting public expenditures to 6% of GDP (Jensen, 2018, p. 6). Commodity booms, however, saw increased pressure on the ruling party to deliver tangible dividends before the country's first post-war general elections. This increased public expenditure from 40% in 2003–2005 to 140% in 2006–2008 resulted in significant budget deficits (Jensen, 2018, p. 7). This spending was designed to allow the People's Movement for the Liberation of Angola (MPLA) to enjoy a broader appeal among the voters during the August 31, 2012, elections, as it favored the disbursal of short-term benefits over long-term planning and investment (Amundsen, 2014). This disbursal had little impact on Angola's need for reconstruction in the post-war context, and reports of systemic corruption made potential Western donors leery (Gastrow, 2017).

China stepped in with loans, and Angola initiated a system of oil-backed credit lines mainly, but not exclusively with Chinese financial institutions. Between 2004 and 2007 alone, Chinese line of credits amounted to $6.5 billion (Zhao, 2011). Increased oil production, and credit lines—that swapped infrastructure projects for commodities, also known as the "Angola model"—provided the country with an "unusually strong access to finance" (Jensen, 2018, p. 5), allowing the political elites to embark on massive infrastructure construction. Simultaneously, the Angola model led to an increased power of the President to disburse largesse to his family, cronies, and the top cadre of the ruling party and make the financing opaque (Gastrow, 2017).

The Angola economy is highly dependent on international oil prices and the price volatility during the global financial crisis negatively affected the country's GDP growth, which nosedived from 13.8% in 2008 to 2.4% in 2009, due to the collapse of oil prices from $145 per barrel in July 2008 to all-time-low $30 in December 2009 (Vines & Weimer, 2011, p. 8). The country's very narrow non-oil tax base and a nonexistent manufacturing sector limited Angola's ability to withstand the collapse of commodity prices. For example, as Vines and Weimer show, in 2008, the non-oil tax only increased by 1.1% of GDP, from 7.3 to 8.4% (2011, p. 8). Simultaneously, the country produced minimally and imported close to 100% of its machinery, transportation equipment, and manufactured luxury goods and a large proportion of its food (2012, p. 9). The combustible mix of volatile commodity prices and exchange rates created a vicious cycle of reducing trade flows and investments. Even though Angola saw its share of oil revenue rise, the boom and bust created considerable uncertainty and discouraged investments. The IMF (2009) reported that the concordant food, fuel, and finance shocks increased the vulnerability of countries like Angola and affected their ability to respond to the crisis.

The significant dip in the oil price in 2008–2009 resulted in substantial economic instability, which pushed the Angolan Government to seek IMF support in November 2009. As a policy measure to mitigate the highly procyclical nature of public investments in Angola, the IMF recommended, as part of its Stand-By Arrangement (SBA), for the Angolan government to create a sovereign wealth fund/oil fund (Jensen, 2018), that commits to "a medium-term fiscal framework to its spending" (IMF, 2011, p. 8)—to separate fiscal expenditure from short-term fluctuations in the prices of oil. The utmost goal was to promote Angolans' social and economic improvement and enhance equitable intergenerational transfer (Chen, 2019 see AfDB, 2009, p. 194). The Fund was expected to strengthen the Government's ability to manage its resource rents and oil price volatility. As Mendes and McClelland (2012) reported, the SWF was initiated with 5 billion and "the IMF promoted setting up a fund and cutting quasi-fiscal operations of the state oil company Sonangol when it loaned the nation 1.3 billion after crude prices fell in 2008" (Mendes & McClelland, 2012).

However, the enthusiasm that an Angolan wealth fund could allow for more efficient management of natural resources became a highly contested issue. The abysmal failure of the ubiquitous state-owned oil company and the second largest in Africa—Sonangol—to translate earnings from hydrocarbons to sustainable economic growth, cast aspersions on whether sovereign wealth fund could perform differently. In 2011, the IMF expressed grave concerns about "large residual financing items" of about $32 billion from 2007 to 2010 in the General State Budget (IMF, 2011). Many political opponents and human rights groups criticized the proposed Fund's setting due to "its secretive finances and super-rich political elite" (McGroarty, 2013). Nevertheless, while many others were not entirely opposed to a wealth fund, their resistance was toward the choice of the former President's son to oversee the day-to-day running of the Fund, as the article in The Wall Street Journal aptly captured, *Angola Wealth Fund is Family Affair* (McGroarty, 2013). However, the lack of debate and public consultation shows how difficult it was for lone voices and political opponents to shape a sovereign wealth fund's future or even push for its well-integration into the country's fiscal framework.

Governing the Angola Fund

The Angola Fund has been governed under Presidential Decree 48/11, art 1 (2) of June 19, 2013, which statutorily vests the Fund with the status of a separate legal personality: "The Sovereign Fund of Angola is a legal entity with legal personality, with administrative, financial and patrimonial autonomy" (IFSWF, 2019) that ensures it.

(a) manages its budgets and revenues; and
(b) practices act independently from the State without prejudice to the powers of supervision and oversight of the President, and that its capital originates from the General State Budget (FSDEA, 2013).

In contrast to countries such as Botswana, Chile, and Ghana, whose funds are managed by the central bank, the mandate of the Angola Fund is different from that of the *Banco Nacional de Angola*. It has direct supervision from the President, as specified in Articles 5, 9, & 26 of Presidential Decree No. 48/11 (March 9) and in Law No. 02/13 (March 7)—enacted by the Parliament through the General State Budget (FSDEA, 2013).

Presidential Decree 48/11 also designated the Angolan government as the sole shareholder of the Fund. The former president, José Eduardo dos Santos, established the oil for infrastructure fund (*Fundo Petrolifera de Angola*) through the presidential decree 48/11 (September 3), which metamorphosed into the Fund (FSDEA, 2013). The Angola Fund's distinct legal personality enhances the operational independence on the part of managers and is intended to insulate the Fund during regime changes. Recent regulatory changes that have sought to strengthen the Fund's governance structure have retained the Fund's separate legal status.

The legal scaffold of the Fund is made up of three presidential decrees. The first allows the Fund to redirect resources from the oil fund account and specifies its governance principles and organizational structure. This enactment mentions the seed capital of $5 billion, and fiscal rules for saving the value of 100,000 oil barrels per day and withdrawals of 7.5% of assets under management for social projects and up to 20% of liquid assets available for emergency (NEITI, 2017, p. 15). The second provides three boards—administrative, advisory, and audit—to run the Fund. The third defines three policy and management rules for the Fund: capital preservation, maximization of long-term returns, and infrastructural development (IMF, 2014).

As defined in the Presidential Decree No. 212/19 of July 15, the Fund's governance structure specifies three bodies: The Board of Directors, the Fiscal Council, and the External Auditors. A five-member Board of Directors leads the Fund; this board has been chaired by Armando Manuel in 2012, José Filomeno dos Santos in 2013, and Carlos Alberto Lopes in 2018. The President appoints members of the board of directors for a five-year term, renewable once. According to Article 10 of the Organic Statute, the board is "responsible for all actions required for managing the Fund and carrying out its duties" (IFSWF, 2019). The board of directors' responsibility is to ensure that the Fund substantially complies with internal policies and procedures, and international best practices such as the Santiago Principles. In carrying out its duties, the Board of Directors is statutorily vested with powers to, among other things, define the objectives, the strategy, and the policies of management of the fund; formulate administrative policies and regulations for the internal management of its activities; design a policy of investment and the annual strategy for achieving investment targets; and formulate a Code of Conduct for its activities (Oshionebo, 2015, p. 227).

The design of the Fund aimed to entrench transparency. For example, Article 7 of the Organic Status the Fund, requires the board to provide information about investments and other data about the Fund to public bodies, submit its annual report for the President's approval and the Finance Minister's review. Members of the Board of Directors are prohibited from holding any public office that competes with their roles as board members and must declare their assets to the Public Prosecutor's Office.

The Fiscal Committee is responsible for the internal supervision of the Angola Fund. The committee performs four critical roles in strengthening the funds: ensuring the Fund complies with the rules that regulate its activities; certifying the value of assets of the Fund; examining the financial statements of the Fund and whether the valuation approach adopted leads to a correct assessment of the assets and profits; supervising the assets of the funds; and pointing out any irregularities brought to its attention by relevant authorities (FSDEA, 2013).

The Advisory Council is responsible for presenting the Angolan Government's position as the owner and investor (IMF, 2014) and is chaired by the Minister of Finance. The Council includes the governor of the *Banco Nacional de Angola*, the ministers of planning and territorial development, finance, and economy. This official body counsels the President of Angola on policies, investment approaches, and presents the annual reports. The Fund's accounts are independently and externally audited.

The Presidential Decree No. 213/19 (the Investment Policy) Article 8 empowers the Angola Fund to appoint external managers and implement relevant guidelines (IFSWF, 2019). As argued later in the paper, both the appointment of the external manager and its investment strategies shows how, despite adhering to international best practices and achieving high ranking on transparency indices such as the Linaburg-Maduell Transparency Index or sovereign credit ratings, sovereign wealth funds could still "legally allow crony capitalism without oversight, rather than to encourage prudent, rule-based investments" (Markowitz, 2020, p. 18).

The Angola Fund has internal and external oversight checks, for example, the Audit Committee is appointed by the Minister of Finance, to whom they submit semi-annual reports on the Fund's management. According to Armando Manuel, a former Finance Minister who also chaired the Fund's advisory board, the appointment of the three-member committee in 2013 was to reposition the Angola Fund "to operate sustainably and spur economic growth and wealth creation for the Angolan citizens, which will ultimately guarantee prosperity in our country" (Africa Outlook, n.d.) while adhering to global standards of governing wealth funds. In line with his argument, the former Angolan President, empowered by Article 19 of Decree No. 48/11, appointed an international accounting firm, Deloitte, in November 2013 as an independent auditor to the Fund. Deloitte first audited the Fund's accounts and submitted the reports to the Ministry of Finance and the Board of Directors and published on the Fund's website. Such independent audit mechanisms have sought to enhance the transparency of the funds' operations and increase its accountability to the State and the Angolan citizens.

In 2016, the Angola Fund transited from the country's national accounting standard for financial institutions to the International Financial Reporting Standards, and its results are accounted annually by Deloitte. The Fund had maintained its 8/10 score in the Linaburg-Maduell Transparency Index, even though these calculations are based mainly on self-reports rather than independent external evaluation (*The Economist*, 2017). Many officials, such as *Hugo Gonçalves*, a former member of the board of directors, consider "the positive evaluation of FSDEA by SWFI [as] a testament to its commitment to the highest standards of financial reporting worldwide" (Africa Business Insight, 2017), a sentiment also reported by Backer (2015, p. 17) that the Angola Fund's "organizational structure is exceptional and generally conforms to the Santiago Principles for an autonomous organization under the guidance of the state." Nevertheless, as the revelations from the *Paradise Papers* (Meisel & Grossman, 2017) and the new president's purge have demonstrated, the Angola Fund was, if anything, far from transparent, despite its consistently high ranking by Linaburg-Maduell Transparency Index.

The Objectives of the Angola Fund

The Angola Fund's primary goal is "to promote growth, prosperity, and social and economic development" (IMF, 2019, p. 43) through economic diversification by investing revenues from oil within the country and abroad in a way that maximizes returns and minimizes the risks of overreliance on oil. The Angola Fund's mandate, therefore, goes beyond the prudent management of resource revenues. It also works toward intergenerational equity due to the exhaustive nature of resources and its citizens' social infrastructure. The Presidential Decree No. 89/13 (June 19) assigns the Angola Fund the responsibility of promoting development, maintenance, and management of large infrastructural projects (FSDEA, 2013).

Since its establishment, the Angola Fund has served as a savings fund and a development fund. The updates of the Fund's legal framework via Presidential Decrees No. 212/19 and No. 213/19 have consolidated the Fund's savings and development functions while also setting up a separate fund to carry out stabilization functions. The new focus is to ensure "organizational models of sound governance, with a clear and effective division of functions and responsibilities, compatible with the nature of [its] activities" (IMF, 2019, p. 43) (Table 1).

The move to create a fiscal stabilization fund to complement the Angola Fund's savings and development mandate has been a direct response to criticisms that the Fund should "serve the purposes of stabilization, investment, and intergenerational equity" (AfDB, 2009, p. 194). The lack of a designated stabilization role has led many to question the efficacy of the Fund, considering that many SWFs such as the Nigeria Sovereign Investment Authority are not only building a savings base for

Table 1 Select African natural resource funds (NRFs)

Country	*SWF name*	*Type*	*legal form*	*Governance structure/operational management*
Angola	Fundo Soberano de Angola	Savings fund Development Fund Fiscal stabilization fund	Separate legal personality	Board of Directors
Botswana	Pula fund	Savings fund		Central Bank (i.e., Bank of Botswana)
Ghana	Ghana petroleum funds	Fiscal stabilization fund Savings fund		Central Bank (i.e., Bank of Ghana)
Nigeria	Nigeria sovereign investment authority	Fiscal stabilization fund Savings fund Development fund	Separate legal personality	Board of Directors

Source Oshionebo (2015, pp. 221–222)

the citizenry but also performing stabilization functions during times of economic shocks. The implications for the Angola Fund, as The Economist argued is that "[T]his means it cannot be used to bolster Angola's international reserves when these are under pressure, as they are at present, because of the low price of oil, the country's main export" (*The Economist*, 2017). The recent move toward a fiscal stabilization fund should be applauded. However, as the International Monetary Fund aptly notes, Angola's top priority as a highly indebted oil exporter should be to reduce its external debt to a sustainable level. It notes that.

> When public debt stabilizes at desirable levels, transfers to the stabilization fund and the FSDEA may start according to pre-determined rules. Those may require appropriate asset accumulation to be achieved in the fiscal stabilization, before considering transfers for long-term saving and development objectives. (IMF, 2019, p. 46)

Monk (2013) states that countries like Angola have distinct challenges from those of the developed countries. While the latter may focus on issues of the Dutch Disease and intergenerational transfer, the former is often cash and credit constrained, and may have to deploy some of its funds at home. He recommends a sequence of economic policies where commodity-rich countries first set up a stabilization fund, followed by a development fund to catalyze growth in the domestic economy, specifically in infrastructure, before setting up an intergenerational and offshore savings account.

However, Angola has done the opposite. It first established savings and development funds without a fiscal stabilization component. The stabilization role is handled separately by the Government of Angola via the *Banco Nacional de Angola* as Oil Price Differential Account, the Strategic Financial Oil Reserve Account for Infrastructure, and the Bonus from Oil Concessions (FSDEA, 2013). However, the Fund has express provisions for the Ministry of Finance to withdraw funds in moments of fiscal shocks, with approval from the president, a provision that requires that the Angola Fund mandatorily has a minimum allocation of 20% of liquid assets investments year-round (FSDEA, 2013).

The Fund's development fund aims to invest in social and physical infrastructure is presented as qualitatively different from what such a fund may achieve in the developed countries. For example, José Filomeno dos Santos, then Chairman of the Angola fund, stated during a conference held on Africa's sovereign funds at Chatham House:

> The cases of Angola and Africa will differ from the case of Norway ... For instance, the Norwegian population might not even be aware in their daily lives of any investment being done by their sovereign wealth fund. Their society is not in a situation where it should be impacted by this amount and type of investment. In Africa, the situation is very different. There are social needs and investment needs at various levels. (Quantum Global, 2014, p. 5)

In short, the aim of the Fund was to make concrete and noticeable investments that improved and addressed urgent social needs and invested in infrastructure.

Angola's Fund Investment and Risk Management Framework

Where does the fund invest and how does it handle risk? The Fund's pre-assessment of risk for portfolio investments lines up with three principles: protection of capital, maximization of long-term returns, and development of infrastructure for the benefits of the citizens (FSDEA, 2013). First, the protection of capital using a third of the Fund's investments, through safe instruments and hedging in low risk assets, which may include fixed income and cash instruments in regulated, mature markets and issuance of sovereigns, large corporations and financial institutions as well as other securities issued in regulated markets with substantial levels of liquidity (FSDEA, 2013, p. 24).

The Fund invests one-third of its portfolio to long-term investments that can generate higher returns and develop infrastructure. The target here is on "commercial infrastructure, agriculture, mining, private equity, real estate, and other allocations in emerging markets, focusing on Angola and sub-Saharan Africa" (FSDEA, 2013, p. 35). In infrastructure, the Fund has invested $1.1 billion into energy, transport, and industrial development (Quantum Global, 2017). In January 2017, the Fund announced that it would invest $180 million into Cabinda's deep-sea port. The amount will complement the $600 million sourced from China EXIM lending to complete the port project (Quantum Global, 2017). Besides, the Angola Fund has invested in agroindustry ventures and infrastructure.

The Angola Fund is one of the three sovereign wealth funds on the African continent (the others are Gabon and Morocco) with an explicit regional mandate (Konfidants, 2018) to focus on high-risk, high-potential private equity in emerging markets. For example, the Fund's pan-African investment mandate allowed it to deploy its $500 million Hotel Fund and health care and infrastructure sub-funds financing within the African continent (Quantum Global, 2017, p. 9). In terms of its $5 billion target allocation, the Fund allocates 50% of its resources to low-risk assets. The remaining 50% to emerging markets and priority sectors on the continent (Quantum Global, 2017, p. 9). The former chairman of the Angola Fund argued that investing half of the endowed funds in "high-quality cash and fixed-income securities (and) and global equities around the globe [will serve] to mitigate domestic and regional-specifics risks" (Intel, 2014).

The Fund also acts as an "umbrella fund" with multiple mandates and diverse portfolios to generate sustainable financial and social returns. The Fund's Social Charter allocates 7.5% of the funds to social impact projects in education, employment generation, information and communication technologies, vocational and technical education, and off-grid access to water, health care, and energy (IMF, 2014). Target beneficiaries are the youth, women, and veterans, demography that could easily delegitimize the incumbent regime. Social impact projects totaling $22 million received funding in environmental protection areas, which benefit local communities or society at large (FSDEA, 2014, p. 25). For example, in 2015, the Fund established its first venture worth $250 million each for mining, logging, agriculture, and entrepreneurship. It also placed $400 million into health care. It has also invested in small venture capital funds and policies to support start-ups, small and medium enterprises to positively affect the informal sector (Konfidants, 2018), and to promote Sustainable Development Goals (Chen, 2019).

One-third of the Angola Fund funds are allotted as "opportunistic investments internationally" in restructuring distressed assets, which the Fund could take advantage of and turn the assets around (Wright, 2014), despite their "much riskier investment" opportunity (Markowitz, 2020). At the same time, as Table 2 shows, 20% of the Fund's assets must be in liquid securities, which can never drop below this level. Besides, because the Fund is primarily funded through oil revenues, investments in the oil sector are not expected to exceed 5% of assets under management (FSDEA, 2013).

The investment strategy has recently shifted under President João Lourenço, which requires the Fund to invest between 20 and 50% of its assets in fixed-income instruments, issued by supranational agencies or institutions of mainly G7 countries; up to 50% in equity securities; and up to 50% in alternative investments (IMF, 2019, p. 44). According to the International Monetary Fund, the risk management approach also "set a limit of a maximum of 30 percent of total assets to be managed by any single asset manager," who must have "at least ten years of investment

Table 2 Investment strategy of the Angola fund

S/No	*Investment strategy*	*Percentage*
1	Private equity	10
2	Agriculture and mining	10
3	Real estate	10
4	Investment in infrastructure	30
5	Investments in distressed asset opportunities	5
6	BRICS and border markets	2.5
7	Commodities	5
8	Social development projects and socially responsible investments	7.5
9	Liquid assets	20
10	Investments in liquid assets and debt	100

Source FSDEA (2013, p. 54)

experience with G7 countries and US$ 3 billion under their management, and be subject to the jurisdiction of a national regulator" (IMF, 2019, p. 44). In many respects, the Fund's investment and risk management are strengthened to guide against a repeat of putting all the Fund's eggs in one basket in which Mr. Bastos was given the responsibility of investing almost all of the Fund's money, instead of hedging the risk.

The Political Entanglements of the Angola Fund

The Angola Fund has consistently been highly ranked on the Linaburg-Maduell Transparency Index as one of the most transparent in Africa, at least until 2017, and the Fund's establishment contributed to an improved credit rating for the country's economy. The rating was revised and upgraded by all three rating agencies following the investment of the Angola Fund in critical infrastructure (Jover et al., 2012). The Fund's focus on promoting economic growth and prosperity resonates with many other commodity-rich countries who also face volatile global prices and revenues, and the Fund may be a way to avoid the resource curse or the paradox of plenty. Despite its strong legal foundation and institutional frameworks built on international governance benchmarks and best practices (Backer, 2015), the Angola Fund is entangled in a series of issues.

We identify three key issues that highlight the challenges of the Angola Fund. First, the performance of any sovereign wealth fund cannot be disentangled from the owners' politics and institutions. What happens when such a country suffers, to echo Botchway, from a "weak regulatory framework for resource exploitation"? (2011, p. 5). The mere existence of a resource fund is not a guarantee of enhanced performance. SWFs must deal with agency challenges relating to the political leaders' undue interference in the daily management of the funds and the extent to which they can represent the citizens (Gilligan et al., 2014). The agency problem becomes compelling if SWFs invest in domestic financial assets, where there are no effective safeguards and the civil society is weak. The combination of a strong President and weak opposition, along with a wealth of exploitable resources produced a specific environment whereby the MPLA government in Angola leveraged oil and diamond rents to hamstring political opposition, civil society, and independent media. It reinforced the government spending as a source of goodwill for political elites in return for political and electoral prospects (Amundsen, 2014). The use of natural resource rents to favor a select group of people against others does not only tend to affect fiscal policies negatively, but it could turn resource rents into a "price-value" where groups may fiercely jostle for control over the state (Fearon & Laitin, 2003). Post-war Angola's political and economic environment shows how the almost sole reliance on oils—linked to the concentration of power in the presidency and the lack of a virile opposition has stymied efforts to manage its natural resources.

Second, while a robust legal, institutional, and fiscal framework is essential, so are the dramatis personae who operate these funds, as the Angola Fund demonstrates. The Santiago Principles (GAA 9) require operational management to independently implement the sovereign wealth fund's strategy and clearly defined responsibilities. Although the Presidential Decrees No. 212/19 and No, 213/19 have provided a new legal framework for the Fund, the ultimate oversight of the "President of the Republic" as shown in Table 3 has continued to raise considerable concerns. Despite the existence of supposedly independent bodies and parliamentary oversight exercised through the budget, the Fund functions as an appendage of the presidency. The Angola Fund can only achieve its mandates where there is a robust system of governance and disclosure insulated from political interference.

Table 3 Approving authorities of NRFs in sub-Saharan Africa

Country	*Purpose*	*Year est*	*Fiscal rule (saving)*	*Fiscal rule (withdrawal)*	*Approving authority*
Angola	Development, fiscal stabilization, savings,	2012	100,000 oil bpd	7.5% for social projects; up to 20% of liquid assets available for emergency	President
Botswana	Stabilization, savings	1994	Budget surplus based on fiscal monetary policy	Budget surplus based on fiscal monetary policy	Automatic trigger
Nigeria	Stabilization, savings, development	2012	Difference between budget and actual oil revenue	Up to 100% of the stabilization component and 60% of SWF profit	National Economic Council

Source NEITI (2017, p. 15)

There are many concerns regarding the President's role as an approving authority in the absence of strong parliamentary oversight. For example, the IMF suggested that the Fund send its periodic report to the legislature and not just the President, as has been the case (IMF, 2014). In Angola, the ultimate oversight of the President has politicized the Fund, and as Chelsea Markowitz recently argued, the Angola's case demonstrates that when a fund is instituted by a politicized entity, the fund objectives, rules, and accountability mechanisms will likely be crafted loosely to allow existing patronage networks to be maintained, even if some protocols are put in place in a bid for international credibility (Markowitz, 2020, p. 19).

Third, the Fund has followed the path of existing financial institutions such as the behemoth National Oil Company, Sonangol Group, and has inevitably become embedded in networks of patronage, cronyism, and the disbursal of resources for short-term benefits without recourse to the long-term development impact. The appointment of José Filomeno dos Santos, the eldest son of the former President, as the Chairman of the Fund in 2013 generated disaffection within and outside Angola. Not only did many civil society organizations and opposition criticize the appointment, but there was apprehension that the move was aimed at "bolstering his son's patronage base, possibly in preparation for him to succeed to the presidency" (Cabeche, 2013). Article 24 of the Presidential Decree No. 48/11 requires that the Chairman of the Fund should be a professional with extensive experience in finance and public management. Sovereign wealth funds such as the Nigerian Sovereign Wealth Fund have appointed officials with considerable investment banking and private equity experience. The opposite is with Angola's Fund, where it was widely believed that Jose Filomeno dos Santos was appointed into the chairmanship position, not because of competency but familial ties. This perception has sent wrong signals across the financial sector and the Angolan citizenry (FSDEA, 2013).

The management of the Fund was not kept at arm's length as the Santiago Principle 9 specifies. The appointment of Quantum Global Investments Africa Management as the sole manager of the $5 billion portfolio, whose chief executive Jean-Claude Bastos de Morais, has extensive ties with José Filomeno dos Santos, has ended in extensive court cases. José Filomeno dos Santos served as a former director of *Banco Kwanza Investe*, established by Mr. Bastos de Morais, and only resigned his position and gave up his shares to assume the position of the Chairman of the Angola Fund to avoid conflict of interests (Pegg, 2017). In addition, Ernst Welteke, the chairman of *Banco Kwanza Investe*, sat on the advisory board of Quantum Global. Even though José Filomeno dos Santos claimed that Quantum Global was selected from a pool of many applicants due to its competences and the Fund would be hiring more managers that never happened (Markowitz, 2020). These ties underscore the dynamic relations between politics and finance in Angola (Saigal, 2014).

The entanglements come together with these three issues, whereby the Fund, government institutions, and the key actors subverted the aims of the fund for potentially personal gain. The Paradise Papers showed how Mr. Bastos de Morais used the Fund to invest in a company he created to construct deep seaports in Cabinda without a competitive public tender (Pegg, 2017). In another revelation, a company owned by Mr. Bastos de Morais, *Afrique Imo Corporation*, was awarded a contract to build a hotel, an office, and a retail complex despite the clear conflict of interest. De Morais not only argued that these investments were aligned with the share holder's interests and the sustainability of the funds, but equally countered reports that showed investing the funds in hotel projects were not economically viable (Africa Center for Energy Policy, 2017).

Concerns over the Fund's operations arose despite the oversights and checks and balances put in place. The Fund's investment and risk management frameworks allowed the funds to be invested in projects based on political relevance (Markowitz, 2020) and could enrich a particular individual or groups, which may explain why Mr. Bastos de Morais was cleared of all charges, even as the $90 million fees paid to Quantum Global and their offshore management continues to generate so much controversy (Africa Center for Energy Policy, 2017). Article 4 of the Presidential Decree No. 89/3, allows the Fund to invest in domestic assets without any budgetary approval, and the manager can choose to invest any amount in any asset class.

The charges brought against Mr. Bastos de Morais, were recently dropped by the attorney general, in light of new evidence brought before a London UK court regarding an undisclosed addendum of the contract between the Angolan Government and Quantum Global (Africa Risk Consulting, 2019). As the director of the National Asset Recovery Service, Eduardo Rodrigues, concluded, the Quantum Global Investments Africa Management had legitimacy and legality to manage the assets of the Sovereign Fund, thanks to a contract concluded in light of

English law with this institution of the Angolan State, despite [its] being extremely damaging to the Angolan State (Africa Risk Consulting, 2019).

Moreover, the recently disclosed addendum is a clear example of how the Fund loosely crafted its accountability mechanisms to benefit private interests. President João Lourenço admitted to this legal gaffe and the implications for transparency and accountability when he argued that:

> Against all the norms and rules of operation of the State Sovereign Funds, in this particular case of Angola, for reasons still not understood, the funds were run by a single foreign entity that the FSDEA itself did not control. (Africa Risk Consulting, 2019)

President João Lourenço has recovered $3.35 billion of assets under the management of Quantum Global Investment Management (Mendes, 2019). In what smacks of a negotiated settlement for the return of all assets under the management of Quantum Global, the alleged scandal, corruption, and conflict of interests at the Angola Fund have almost been brought to a final close (Mendes, 2019).

The implications of the political entanglements, especially with the power located in the president's office and issue of opacity this raises affect some of the most innovative aspects of the Fund such as the Fund's domestic infrastructural investment focus and the Social Charter which aimed at improving the people's well-being, and the protection of the environment. For example, financing domestic infrastructure projects such as clean energy, water supply networks, agriculture could raise issues of transparency and risk and may undermine public financial systems (IMF, 2019) if such investments are government spending operations that could be achieved through direct budget financing and implemented by ministries. Questions regarding the coordination of the Fund with the budget, in its role for development purposes such as supporting foreign direct investments, therefore needs to be addressed to avoid poor investment decisions, conflict of interest, patronage, and corruption.

The structure of the fund presents challenges, specifically, the Presidential Directive 107 which has defined the Angola Fund's investment policy, namely, capital preservation, maximization of long-term returns, and infrastructure development. This means the Fund is both a development and savings fund (IMF, 2014). The saving mandate has often been subjected to criticisms. As Elias Isaac, the director of the Angola office of the Open Society Initiative in South Africa, quizzed, "[B]ut how can you even think about that when such a large portion of the current generation is living in poverty without access to basic services"? (Fahamu, 2015). Its initial objective of savings has changed over time to include areas such as the hospitality industry, although there has not been any explicit mention of this change. Many have cautioned that such "mutability of mandates" could make it difficult to assess the Fund's performance (Fahamu, 2015).

Recently, the Fund has had a fundamental change. President João Lourenço promised to strengthen it but his decision to withdraw a whopping $2 billion from the Fund to finance rural infrastructure projects raises concerns regarding not only the purpose of the Fund but also its supposedly autonomy. In August 2019, his government withdrew $1 billion for health, water, education, health, sanitation, and road infrastructure projects (IMF, 2019, p. 43).

Many have questioned why such investments should be financed through withdrawal from the Fund rather than budgetary provisions. And here the entanglements come into focus again. The President can choose to bypass the Parliament and disburse resources for short-term benefits to his allies and constituents. Such development allows for "reduced market discipline on public projects, weaker regulatory oversight, and diminished scrutiny from parliament and other budgetary checks and balances" (Gilligan et al., 2014, p.9). That half of the Fund can be withdrawn at the President's beck and call raises concern regarding the funds' autonomy. Moreover, as Markowitz argues, "this is an example of how a heavily politicized fund can easily be delegitimized after regime changes and raises questions over whether the fund should have been established in the first place" (2020, p. 20).

The Angola Fund's record toward diversification remains a distant dream due to the lack of transparency in its investment portfolio and volatility in the global price of oil. Four areas better qualify this failure. First, the Angola economy is monolithic and continues to depend on revenues from commodities despite the vagaries of global oil and diamond prices and exogenous shocks, as 90% of total exports in 2018 were from oil and gas (OPEC, 2019) while importing virtually all essential commodities and food (Vines & Weimer, 2011). Second, the Fund's failure to hedge risk by giving Jean-Claude Bastos de Morais the responsibility for investing almost all its funds showed its penchant for non-diversification. Third, despite the rules of supplementary funding of 100,000 barrels per day, no subsequent funding has taken place since its inception (IMF, 2019, p. 43). Fourth, while the Fund was to invest foreign reserves and finance economic diversification in 2012, it only received the final part of its $5 billion endowments in 2014 (Jablonski, 2015). The late financing made it extremely difficult for the Fund to adapt to falling oil prices, which were as down by 40% by June 2015 (Jablonski, 2015).

The Fund's decision to invest in non-oil and gas assets in Angola and across Africa is commended. The target was to ensure the investments do not correlate with revenues due to Angola's exclusive dependence on oil and the associated risks of fluctuating commodity prices. Nevertheless, investing a significant portion of investment in infrastructure and domestic tourism has not helped diversify the economy as essential goods continue to be imported (Vines & Weimer, 2011). The falling oil price—resulting in low revenues and corresponding low savings and increased public debt—demonstrates the Angola Fund's failure to promote diversification almost ten years after its creation.

Conclusion

Recent analyses of sovereign wealth funds make an implicitly deterministic argument that adherence to rules governing sovereign wealth funds could allow the Fund to achieve fiscal stabilization, macroeconomic, and developmental goals. What happens in a country like Angola with a "weak regulatory framework for resource exploitation?" SWFs' political economy allows for an explanation of these funds' practices that may diverge with the rulebooks. The focus on domestic and regional mandates where regulatory institutions, governance structures, and domestic oversights are weak, for example, has allowed for financing projects in which the President and his allies have vested interests. For example, the Paradise Papers revealed how Mr. Bastos de Morais used money from the Angola Fund to invest in companies he set up. The Fund's goal is to "substantiate these investments" and not ensure they improve citizens' lives or diversify the economy.

President João Lourenço's 2017 election as the ruling party leader is regarded by many as a "shake-up of an older order" (Onishi, 2018). Since President João Lourenço took over from José Eduardo dos Santos, there has been a shift toward improving governance, fighting corruption, and creating a business-friendly environment (IMF, 2019). Many regard his focus on reforms as "an opportunity for forging a new social contract" that would do away with unequal power relations that have for a long time empowered the President and his camp followers while depriving Angolans the opportunity to participate in governance (World Bank, 2018, p. 56). He has embarked on a series of regulatory reforms in the extractive sector. In March 2018, he removed and charged José Filomeno dos Santos as the chairman of the Fund and replaced him with Carlos Alberto Lopes. President João Lourenço also removed Isabel dos Santos as the head of Sonangol Group, dismissed Quantum Global as the Fund's manager, and sued Mr. Bastos de Morais; the asset manager won the case, with the state spending up to $10 million as litigation fees. His inability to convict Mr. Bastos de Morais, the enormous amount spent on the case, and what seems like a negotiated settlement that has led to the almost final closure of the case further diminished the Fund's credibility and affected its performance (Markowitz, 2020).

In addition, while there was initial hope that President Lourenço could reorient the Fund to serve the public interest, there are strong concerns that the Fund may not achieve "operational independence from the owner" (IMF, 2014, p. 46) as the recent decision to withdraw $2 billion out of the remaining $3.5 billion total assets under management demonstrates. Certainly, such a move will reduce external loans, but it could also raise transparency and accountability issues since the approving authority is the President of the Republic.

Finally, the Angola Fund's target of infrastructural development is innovative, considering the difficulties of raising financing from the global capital market. External sources of long-term finance for infrastructure in sub-Saharan Africa have diminished as the continent's financial markets grapple with low liquidity (Arezki & Sy, 2016; Hove, 2016). However, while we argue that the Angola Fund's innovation in investing in social infrastructure could close the financing gaps for Sustainable Development Goals, this can only be realized in a depoliticized legal, institutional, and fiscal framework, where development functions are delineated with extensive checks on the authorizing figure of the President.

References

AfDB. (2009). *Oil and gas in Africa*. Oxford University Press.

Africa Business Insight. (2017). Fundo Soberano de Angola retains high ranking on the sovereign wealth fund institute transparency index. *Africa Business Insight*, https://www.howwemadeitinafrica.com/fundo-soberano-de-angola-fsdea-retains-high-ranking-sovereign wealth fund-institute-swfi-transparency-index/59280/. Accessed 7 August 2019.

Africa Center for Energy Policy. (2017). Rolling back the gains of upstream open contracting: The case of revenue mismanagement. https://s3.amazonaws.com/new-acep-static1/reports/ROLLING+BACK+THE+GAINS+OF+OPEN+CONTRACTING+2.pdf. Accessed 3 September 2020.

Africa Outlook. (n.d.). Fundo Soberano de Angola confirms the appointment of Audit Board and appoints primary custodian. https://www.africaoutlookmag.com/industry-insights/article/462-fundo-soberano-de-angola-confirms-appointment-of-audit-board-and-appoints-primary-custodian. Accessed 7 May 2019.

Africa Risk Consulting. (2019, April 15). Angola monthly briefing. *Africa risk consulting*, https://www.africariskconsulting.com/arc-briefing-Angola-April-2019.html, Accessed 28 August 2020.

Amundsen, I. (2014). Drowning oil: Angola's institutions and the 'resource curse'. *Comparative Politics, 46*(2), 169–189.

Arezki, R., & Sy, A. (2016). Financing Africa's infrastructure deficit: From development banking to long-term investing. *Journal of African Economies, 25*(2), ii59–ii73.

Backer, L. C. (2015). International financial institutions (IFIs) and sovereign wealth funds (SWFs) as instruments to combat corruption and enhance fiscal discipline in developing states. *International Review of Law*. https://doi.org/10.5339/irl.2015.swf.5.

Botchway, F. N. (2011). *Natural resource investment and Africa's development*. Edward Elgar.

Business Monitor International. (2015). Angola operational risk report first quarter. BMI's Industry Report & Index Series.

Cabeche, A. (2013, November 14). Angola: President's son dips into sovereign fund for 'trophy' Savile Row office. https://mg.co.za/article/2013-11-14presidents-son-dips-into-sovereign-fund-for-trophy-savile-row-office/. Accessed 7 December 2019.

CDP. (2018, April). Lessons learned in developing productive capacity: Fourteen case studies. *Committee for Development Policy (CDP) Background Paper* No. 37.

Chen, J. (2019). Financing the sustainable development goals: The role of African sovereign wealth funds. *International Law and Politics*, *51*, 1259–1292.

Fahamu. (2015, Aug 4). Analysis: Do sovereign wealth funds make sense for Africa? http://www.fahamu.org/ep_articles/analysis-do-sovereign-wealth-funds-make-sense-for-africa/. Accessed 7 May 2019.

Fearon, D. J., & Laitin, D. D. (2003). Ethnicity, insurgency, and civil war. *The American Political Science Review*, *97*(1), 75–90.

FSDEA. (2013). *Nurturing our common future – 2013 Annual Reports*. Lunda: Fundo Soberano de Angola.

FSDEA. (2014). *2014 annual report*. Fundo Soberano de Angola.

Gastrow, C. (2017). Aesthetic dissent: Urban redevelopment and political belonging in Luanda, Angola. *Antipode, 49*(2), 377–396.

Gilligan, G., O'Brien, J., & Bowman, M. (2014, May). *Sovereign wealth funds: The good guy investment actors?* (Centre for International Finance and Regulation Research Working Paper Series No. 021/2014).

Hove, S. (2016, April). *Sovereign wealth funds and infrastructure development in Africa* (Quantum Global Research Lab, Working Paper No. 2016/02).

IFSWF. (2019). Angola: Santiago principles self-assessment. https://www.ifswf.org/printpdf/node/2721, Accessed 28 August 2020.

IMF. (2009). *Fiscal rules: Anchoring expectations for sustainable public finances*. Washington, DC: The International Monetary Fund.

IMF. (2011, December). Angola: International Monetary Fund (IMF) country report No. 11/346.

IMF. (2014). Angola: Second post-program monitoring press release and statement by the executive director of Angola. IMF Country Report No. 14/81.

IMF. (2019). Angola. International Monetary Fund Country Report No. 19/371.

Intel. (2014, April 9). Reuters summit – Angolan wealth fund fixed income but no project investments yet. *Reuters*. https://www.reuters.com/article/africa-summit/reuters-summit-angolan-wealth-fund-buys-fixed-income-but-no-project-investments-yet-idUSL6N0N12TI20140409. Accessed 13 June 2019.

IWG. (2008, October). Sovereign wealth funds: Generally accepted principles and practices. "Santiago Principles". International Working Group of Sovereign Wealth Funds.

Jablonski, A. (2015, January 4). Will Angola's sovereign wealth fund allow it to adapt to falling oil prices? *Global Risk Insights*. https://globalriskinsights.com/2015/01/will-angolas-sovereign-wealth-fund-allow-adapt-falling-oil-prices/. Accessed 12 July 2020.

Jensen, S. K. (2018, September). Angola's infrastructure ambitions through booms and busts: Policy, governance and reform. *Africa Program Research Paper*.

Jover, E., Pinto, A., L., & Marchand, A. (2012). *Angola private sector country profile*. African Development Bank.

Konfidants. (2018). *African sovereign wealth funds index 2018*. Konfidants.

Le Billion, P. (2001). Angola's political economy of war: The role of oil and diamonds, 1975-2000. *African Affairs, 100*(398), 55–80.

Markowitz, C. (2020, January). *Sovereign wealth funds in Africa: Taking stock and looking forward*. South Africa Institute of International Affairs (Occasional Paper 304).

McGroarty, P. (2013, February 26). Angola wealth fund is family affair. *The Wall Street Journal*. https://www.wsj.com/articles/SB10001424127887323864304578318683374319560. Accessed 7 May 2019.

Meisel, A., & Grossman, D. (2017, November). Paradise papers: Tycoon made $41m from 'people's fund.' *BBC News*. https://www.bbc.co.uk/news/world-africa-41906123.

Mendes, C. (2019, March 23). Angola says it recovers $3.35 billion of assets from Quantum. *Bnn Bloomberg*. https://www.bnnbloomberg.ca/angola-says-it-recovers-3-35-billion-of-assets-from-quantum-1.1233521. Accessed 28 August 2020.

Mendes, C., & McClelland, C. (2012, October 17). Angola starts sovereign wealth fund with $5 billion. *Bloomberg*. https://www.bloomberg.com/news/articles/2012-10-17/angola-starts-sovereign-wealthfund-with-5-billion. Accessed 7 August 2019.

Monk, A. (2013, February 28). Africa will soon be the global leader in SWFs. *Institutional Investor*. https://www.institutionalinvestor.com/article/b14zbbt3yq91my/africa-will-soon-be-the-global-leader-in-swfs.

NEITI. (2017). The case for a robust oil saving fund for Nigeria. *Nigeria Extractive Industry Transparency Initiative* (NEITI) Occasional Paper Series.

Onishi, N. (2018, September 25). Angola hold ex-ruler's son on fraud charge. *The New York Times*. https://www.nytimes.com/2018/09/25/world/africa/angola-corruption-dos-santos.html. Accessed 7 August 2019.

OPEC. (2019). *Angola facts and figures*. https://www.opec.org/opec_web/en/about_us/147.htm. Accessed 7 August 2019.

Oshionebo, E. (2015). Managing resource revenues: Sovereign wealth funds in developing countries. *ASPER Review of International Business and Trade, XV*, 217 264.

Pegg, D. (2017, November 7). Angola sovereign wealth fund's manager used its cash for his own Projects. *Guardian*. https://www.theguardian.com/world/2017/nov/07/angola-sovereign-wealth-fund-jean-claude-bastos-de-morais-paradise-papers.

Quantum Global. (2014, September 5). *Africa's sovereign wealth funds: Demand, development and delivery. Africa Program Conference Summary*. The Royal Institute of International Affairs.

Quantum Global. (2017). *Sovereign wealth funds as a driver of African development*. Quantum Global Group. http://quantumglobalgroup.com/wp-content/uploads/2017/10/Sovereign-Wealth-Funds-as-a-driver-of-African-development.pdf.

Ross, M. (2015). What have we learned about the resource curse? *Annual Review of Political Science, 18*, 239–259.

Saigal, K. (2014, September 9). Angola's sovereign wealth fund sets sights on alternatives. *Euro Money*. https://www.euromoney.com/article/b12kkj4rndwc5w/angolas-sovereign-wealth-fund-sets-sights-on-alternatives. Accessed 7 September 2019.

Sovereign Wealth Fund Institute (2021) Top 100 largest sovereign wealth fund rankings by total assets. https://www.swfinstitute.org/fund-rankings/sovereign-wealth-fund

The Economist. (2017, November 22). *Angola: FSDEA fees questioned*. http://country.eiu.com/article.aspx?articleid=286148612. Accessed 7 May 2019.

Ufimtseva, A. (2019). Governing extractive industries: The case of Angola's petroleum sector. *Nokoko, 7*, 113–146.

Vines, A., & Weimer, M. (2011, June). Angola: Assessing risks to stability. A Report of the Center for Strategic and International Studies (CSIS) Africa Program.

Wiig, A., & Kolstad, I. (2013). Assigned corporate social responsibility in a rentier state: The case of Angola. In P. Lujala & S. A. Rustad (Eds.), *High-value natural resources and post-conflict peace building* (pp. 147–154). Earthscan.

World Bank. (2018, December). Angola: Systematic country diagnostic. Creating assets for the poor.

Wright, C. (2014, September 5). *Inside Angola's sovereign wealth fund*. https://www.forbes.com/sites/chriswright/2014/09/05/inside-angolas-sovereign-wealth-fund/#33cb23eb8b98. Accessed 4 May 2019.

Zhao, S. (2011, May 25). The China-Angola partnership: A case of China's oil in Africa. *China Briefing*. https://www.china-briefing.com/news/the-china-angola-partnership-a-case-study-of-chinas-oil-relationships-with-african-nations/. Accessed 28 August 2020.

Botswana's Natural Resource Fund (The Pula Fund)

David Sebudubudu

Introduction

Unlike many other resource-endowed African nations, Botswana has never faced a consequential challenge of who owns and controls its natural resource wealth. The 1967 Mines and Minerals Act ceded mineral resources from private owners (individuals and communities) to the state shortly after independence. By surrendering mineral ownership rights to the state, Botswana declared minerals a strategic resource. This decision was achieved following consultation with various communities (Sebudubudu and Molutsi 2011; Sebudubudu and Mooketsane 2016). Botswana's use of natural resources has not been primarily associated with ethnic conflict or divisive politics as it has elsewhere in Africa, where minerals have brought more pain than joy. This article contends that a political culture characterized by democracy and stability has ensured the successful and peaceful development of natural resources in Botswana. The establishment of the Pula Fund, a sovereign wealth fund that recognized the inevitable depletion of resources such as diamonds, was central to this. This article argues that the creation of a sovereign wealth fund is

D. Sebudubudu (✉)
University of Botswana, Gaborone, Botswana
e-mail: Sebudubudu@ub.ac.bw

E. Okpanachi and R. Chowdhari Tremblay (eds.), *The Political Economy of Natural Resource Funds*, International Political Economy Series,
https://doi.org/10.1007/978-3-030-78251-1_4

in line with Botswana's vision of ensuring that its resources are used for the benefit of all its citizens as opposed to regional, sectional, or sectarian interests.

In terms of a number of governance indices, Botswana is one of the few well-governed and successful countries in Africa, an anomaly given the deplorable governance patterns associated with Sub-Saharan Africa (SSA). Indisputably, its success contrasts with many of its resource-endowed African peers that became decidedly corrupt or embroiled in conflict due to abundance of resources. Angola, Nigeria, Sierra Leone, and the Democratic Republic of the Congo are examples of African countries that have been caught in this trap and have thus failed to realize significant socioeconomic and political transition despite the resources they enjoy Today, Botswana is widely regarded as a country that has transitioned from one of the world's poorest countries at its independence into an upper middle-income country, as per World Bank categorization (World Bank 2019). Its success is attributed to its cautious approach toward the use of its resources, particularly diamonds, following their discovery immediately after independence. Alongside this, Botswana has been generally accepted and consistently recognized as one of the few countries in SSA to be afflicted only minimally by corruption (Transparency International 2019), despite some corruption scandals being reported since the early 1990s. The most recent cases of corruption that have been primarily and widely reported on by the privately owned media, while the state media remained mute, include the National Petroleum Fund (NPF) scandal, persistent allegations of corruption against former and founding Director General of the Directorate of Intelligence and Security Services (DISS) Isaac Kgosi, allegations of corruption in the BPOPF against Carter Morupisi, head of the public service, and others, and other cases that implicated a number of ministers in Ian Khama's government. President Mokgweetsi Masisi and former President Ian Khama were also implicated in the NPF scandal, yet they denied such allegations (Sebudubudu 2019). Some of these cases, such as the NPF case, were dismissed on technicalities, while others were in motion or were yet to be registered in court, such as that of Isaac Kgosi, at the time of this writing. These examples seem to suggest a reluctance to prosecute high-profile cases and a tendency to underreport perceptions of corruption, so that reports such as those of Transparency International (2019) do not adequately reflect the reality. However, citizens' perceptions of corruption are worryingly high. For instance, the Afrobarometer survey of 2019 put citizen's perceptions of corruption at 69%, suggesting that citizens consider corruption a matter of public concern.

An abundant and established literature devotes serious consideration to natural resource endowments and the attendant challenges they pose to such countries. The prevailing perspective in the literature correlates natural resource endowment with what has come to be widely known as the "resource curse" (Coutinho 2011). This term refers to the adverse consequences generally identified in most countries endowed with natural resources that hinder rather than facilitate development (Coutinho 2011). This chapter does not seek to reignite the "resource curse" debate, which falls outside its scope. However, it does attempt to explain Botswana's Pula Fund by seeking to understand decision-making within the Pula Fund and the influence of politics on the fund by examining the context under which it was established, its establishment stage, the fund in operation and its challenges, and the political and economic outcomes that were realized.

Botswana: Pre-Establishment Stage

Botswana, which is a functional democracy by African standards, is among those countries considered to have obscure natural resource funds, which puts it in the same league as countries such as Equatorial Guinea, Kuwait, Iran, Mexico, Russia, and Qatar (The Natural Resource Governance Institute 2014). All of these countries have embraced the Santiago Principles, which advance sound governance in the management of natural resource funds (The Natural Resource Governance Institute 2014). Most of the aforementioned countries have no known history of democracy, but have traditionally experienced dictatorship or authoritarian rule. The lack of good fund governance is what in part separates countries that misuse their resources from those that ensure prudent resource management. However, because SWFs are considered to be at the nexus of politics and finance, the lack of a clear demarcation between politics and finance gives rise to "concerns that SWFs will be used illegitimately to advance political, as opposed to commercial agendas" (Ahmadov et al. 2011, p. 42). NRFs or SWFs were, in general, created under this context and for the reasons advanced above.

Botswana, a former British protectorate, has sustained its modest democracy since its pre-independence elections in 1965 and political self-determination the following year. At that time, the country was largely surrounded by adversarial neighbors that were ruled by minority regimes—Zimbabwe (formerly Southern Rhodesia), Namibia (formerly South West Africa), and South Africa, until their respective independence in 1980 and 1990 and non-racial multiparty elections in 1994. Botswana's geopolitical environment called for a cautious and practical foreign policy with its immediate neighbors in order for it to function as a viable state and to gains access to seaports as a landlocked country. A practical foreign policy was especially significant, considering the country's poverty situation at independence and because nationals and foreigners alike doubted its political viability. Charles King of the *Southern Africa News Service of Canada* described Botswana's extreme poverty on September 28, 1966:

> Bechuanaland [Its name before independence]: An impoverished, arid and hungry land without hope of achieving economic stability makes its debut this week among the community of nations. The new blue, white and black flags are flying everywhere in Gaborone, its incongruous capital city. But elsewhere in the vast, trackless wasteland that will take the name of Botswana, there is little to celebrate. Two years of disastrous drought and crop failure have brought havoc and hunger to its widely scattered agricultural inhabitants. More than one fifth of the population is literally being kept alive by emergency feeding and the numbers are rapidly increasing.

King's description illustrates the grim future of surviving as a country that Botswana faced. Fortunately, Botswana discovered large deposits of diamonds in 1967, a few months into independence. This discovery, aided by the political context, played a pivotal role in changing its deplorable economic situation. The Botswana Democratic Party (BDP), which has won all of the 12 multiparty elections held so far, including the most recent on October 23, 2019, assumed a moderate policy stance in 1966 (Sebudubudu and Molutsi 2011), and this made the country friendly to needed foreign investors in light of its lack of know-how to extract its natural resources. The reasonable or moderate policy standing pursued by the ruling party and government allowed Botswana to partner with international capital in its mining of diamonds and other minerals. Minerals, especially diamonds, became the main export and greatest revenue earner for Botswana. For instance, in terms of the OECD (2013), diamonds constituted 84.8% of Botswana's exports in 2012. Minerals did not become an affront to its developmental democracy prospects, as was the case in most African countries. Sebudubudu and Mooketsane (2016,

p. 89), however, caution that Botswana's positive usage of resources should not be seen as "a one-dimensional story of minerals," as minerals are not the only explanatory power behind Botswana's success.

Botswana's accomplishments suggest that in addition to minerals, other insightful considerations or explanations, related to its political culture, catapulted it to success and consequently made it an outlier in Sub-Saharan Africa (see, among others, Leith 2005; Lewin 2011). Leith (2005) locates Botswana's success in its political system that ensured effective leadership and institutions, and Lewin (2011) attributes its prosperity to sound governance, policies, and some luck. Sebudubudu and Mooketsane (2016) relate its success to leadership, tolerance, and effective management of various ethnic groups, advancement of meritocracy in the country's public service, and building a working partnership with the private sector. All of these factors together ensured that the country's institutions and stable political system were somewhat productive and reliable. The stability of the BDP can also be attributed to the fragmented opposition, owing to a number of challenges that work against it, such as lack of resources, failure to convert the opposition into an accepted institution and failure to offer the electorate "with a discernable programme" (Sebudubudu et al. 2016). Nonetheless, the executive dominates decision-making, and its influence extends to the other branches, especially Parliament, which has been diluted by executive control. Furthermore, organized interests have yet to take shape, owing in part to the country's political culture that, among others, promotes deference to authority.

Factors behind Botswana's success may be applicable to its context, but sound leadership (see, for instance, Rotberg 2003; Sebudubudu and Botlhomilwe 2012) stands out in the literature as one of those factors that made a difference to its rapid and much acclaimed transformation. The foregoing factors characterized the political culture in which Botswana created the Pula Fund. Botswana's leaders realized that the country's economy was largely dependent on revenues derived from natural resources that were finite, and in 1994, they made a deliberate decision to establish a Sovereign Wealth Fund, the Pula Fund. Information on the debate over whether or not to establish a fund appears to be sketchy and not readily available, although a few people do seem to be knowledgeable about the establishment of the Fund. The following section discusses the fund's governance and structure, objectives, and management frameworks.

Establishment Stage

Rules and Structures

The Pula Fund, which is considered the largest and longest-running sovereign wealth fund in Africa, came into existence through an Act of Parliament. The Bank of Botswana Act of 1975 set out the terms for the establishment of the Pula Fund in 1994; the Fund was reconstituted in accordance with the revised Bank of Botswana Act of 1996 (Republic of Botswana 1996; Meijia and Castel 2012). Section 35 (1) of the Bank of Botswana Act states:

> If in the Bank's opinion, the primary international reserve is in excess of the amount needed to accomplish its principal objectives and to finance the international transactions of Botswana and is likely to remain in that position for some time, the Bank may, after consultation with the Minister, establish a separate long-term investment fund or funds in which to invest the assets in excess of those needed for the Primary International Reserve.

The Bank of Botswana directly operates and manages the fund, which follows a similar strategy of investment as that of Norway. As it stands, the Pula Fund is not a separate legal entity. It is constituted, in terms of Section 35 of the revised Bank of Botswana Act of 1996, purely as an accounting entity, and forms part of the Bank of Botswana balance sheet.

In accounting terms, the ownership of the Fund is divided into two parts: the government's share and the Bank of Botswana's share. The government has direct access to its share of the Fund only. Although the Fund is a jointly owned entity in terms of institutional infrastructure, the governance of the Fund's assets has been entirely delegated to the Bank of Botswana. The Ministry of Finance and Development Planning, now the Ministry of Finance and Economic Development, has an indirect say or influence on the operation of the Fund (Interview, 26 Aug 2019). 50% of the Pula Fund is managed by the Financial Market Department of the Bank of Botswana while the remaining 50% is managed by nine foreign fund managers (Ntibinyane 2016) appointed by the Bank of Botswana through an undisclosed process. The Fund Managers visit Botswana about twice a year to offer reports on fund performance (Interview with a former Deputy Governor, 3 Oct 2019). The Fund Managers derive their mandate from the Bank of Botswana, which defines assets in which the Managers are permitted to invest. Because the Pula Fund assets form part of the country's foreign exchange reserves, they are invested in foreign currency denominated assets, mainly a mixture of global bonds and equities (Interview, 3 Oct 2019). Because the Fund's investments are all external and are not in Botswana, "the activities related to these investments do not have any significant direct domestic macroeconomic implications" (IFSWF 2019).

Governance of the Pula Fund

The Fund's governance follows a three-tier structure consisting of the Financial Markets Department, Investment Committee, and Bank of Botswana Board (IFSWF 2019). The Financial Markets Department implements the Fund's investment planning and assets allocation, using both internal and external fund managers (IFSWF 2019). The Investment Committee is responsible for decision-making "on the execution of investment strategy, including tactical deviations from the board-approved strategic asset allocation" (IFSWF 2019). Membership of the Investment Committee includes the Bank of Botswana Governor who presides over it and reports to the Bank of Botswana Board, and investment experts from the Financial Market Department (Natural Resource Governance Institute 2013; IFSWF 2019). Furthermore, a former Deputy Governor of the Bank declared that "the Investment Committee does not have any external representatives" (Interview, 26 Aug 2019). Hence, the Ministry of Finance is not involved in this committee, although it is represented on the Bank's Board.

According to Section 10 of the Bank of Botswana Act, the Board consists of the Governor of the Bank, who serves as the chairman; the Deputy Governor, who can attend the Board meetings but has no vote except when representing the Governor as chairman or in his/her own capacity as an appointed member of the Board; and eight other members appointed by the Minister of Finance and Economic Development (MFED). The eight members of the Board include the Permanent Secretary of the Ministry of Finance and Economic Development (MFED), and others, including not more than two public officials, "appointed by the minister from persons of goos standings and experience in business, professional and academic matters" (Bank of Botswana 1996, pp. A.169). The Board decides on the Bank's policies, including the Pula Fund Investment Policy. To this extent, it can be argued that the government is indirectly involved in the running of the Pula Fund through the role of the Permanent Secretary on the Board, and the Board's approval of investment policies. The Pula Fund is part of the external audit of the Bank of Botswana accounts, in line with the Bank of Botswana Act and international auditing standards. It is also worth noting that Botswana participates in the International Forum of Sovereign Wealth Funds (IFSWF) and upholds the 2008 Santiago Principles, which it played a part in drawing, in its management of the Fund (Bank of Botswana 2019).

Objectives of the Pula Fund

There are a number of considerations in the creation of SWFs, which tend to vary according to countries and the Fund at hand. According to Humphreys and Sandbu (2007, p. 197), Natural Resource Funds (NRFs) are normally considered "stabilization funds" or "savings funds." Similarly, Wills et al. (2016, p. ii3) posit that, apart from the promotion of "political accountability, portfolio diversification of investment returns and Dutch disease considerations," the three principal considerations for keeping revenues in SWFs or abroad are "intergenerational transfer, parking and stabilisation." Humphreys and Sandbu (2007, pp. 198–199) contend that NRFs are meant to "facilitate the accumulation of large, volatile, and temporary revenues when times are good; stabilize public spending; and finance public spending when natural resource revenues are no longer flowing in."

The Bank of Botswana's website states that the Pula Fund is a "long-term investment portfolio… with the aim of preserving part of the income from diamond exports for future generations" (Bank of Botswana 2019). Although the Bank of Botswana clearly states that the Pula Fund was set up to save for future generations, other objectives have been identified, and the Fund has been criticized for a lack of clear objectives.

Scholars have identified and commented on various objectives presented in the literature, such as filling the void when the diamond mines are exhausted, creating fiscal flexibility and stability, and serving as a savings account. Ntibinyane (2016), for instance, noted that the Pula Fund was established to secure a portion of diamond proceeds for future generations because diamonds are expected to be depleted in 20 years. The Fund also serves to ensure that foreign reserves were stable. A former governor of the Bank of Botswana declared that the Pula Fund was established as a buffer to build local reserves out of excesses to ensure some stability (Interview, 17 Sept 2019). Meijia and Castel (2012, p. 10) posit that "the objective of the fund was to provide greater flexibility in the management of international reserves and greater certainty in the forecasting of annual dividend payments to the government from the Bank of Botswana." Meijia and Castel (2012, p. 10) further contend that the Fund was set up to serve two main roles: as "a stabilization fund and a savings fund for future generations." The Natural Resource Governance Institute (2013, p. 2) states that "the Pula Fund's policy objective is unclear [as per the 1975 and 1996 Bank of Botswana Acts]; however, it has been used to stabilize revenues and serves as a future generations savings fund." Thus, although the Pula Fund broadly covers all three main considerations for a SWF, a former deputy governor of the Bank of Botswana has also suggested that the objectives of the fund are "ill-defined" because whether it is primarily a stabilization or future generation fund is not clear (Interview, 26 Aug 2019). In theory, the Pula Fund is both a stabilization and future generations fund, but in practice, it has been largely used as a stabilization fund, as seen during the 2008 financial crisis. It can be argued that the absence of clearly defined objectives for the Fund, the lack of clear rules about injections into and withdrawals from the fund, and the size of the fund that would be needed for each of these objectives creates some challenges regarding its sustainability as a future generations fund, particularly in the long term.

Investment and Risk Management Framework

The Pula Fund has some sound investment rules. Specifically, its assets can only be invested outside of Botswana, as prescribed by the Bank of Botswana Act. It collects what remains from any balance of payment surpluses, not the budget, after any allocation of those surpluses to the normal foreign exchange reserves (liquidity fund). It is only the government's share of the Pula Fund that collects budget surpluses. However, the Fund lacks sound deposit and withdrawal rules (NRGI/Columbia Centre 2013, p. 2). Despite Botswana being a member of the Santiago Principles, the Pula Fund only achieved 44/100 in the Santiago Compliance Index, 52/100 in the Resource Governance Index Natural Resource Fund Score, and 56/100 in the Truman Sovereign Wealth Fund Scoreboard. These scores demonstrate deficits in Botswana's governance standards (Natural Resource Governance Institute 2013, p. 2), so that it does not compare favorably with others such as Norway and Qatar (Ntibinyane

2016). Ntibinyane (2016) observes that the Fund has not "been able to live up to many of these [Santiago] principles." Similarly, Botswana has so far failed to put in place firm objectives as discussed above, clear deposit and withdrawal rules, or strong oversight mechanisms to ensure accountability and transparency (Ntibinyane 2016). Consequently, the Fund has been safeguarded by the Bank of Botswana, and its management and performance has been largely left in the hands of the Bank, especially the governor who, as chairman of the Bank's Board, has significant power on the Fund (Ntibinyane 2016). The following section analyzes the fund in operation and the challenges it faces.

Natural Resource Funds (NRFs) in Operation: Political Struggles over the Funds, Continuity, and Change

Although Botswana has been well-endowed with natural resources and, consequently, is heavily dependent on resource revenues for its development trajectory, particularly diamonds, the country has so far not suffered the same fate, that of systemic mismanagement of rents from its resources, as most other SSA countries that are rich in natural resources. As a result, Botswana has been presented in the literature and development circles as a rare case worth emulating. Harvey (2015, p. 1) has noted that "Botswana outperformed every other country in the world in terms of per capita income growth from 1965 to 1998. From 1998 to 2011, gross domestic product (GDP) per capita growth averaged 3.03% a year. The child mortality rate (per 1000 live births) was 52.66; South Africa, by comparison, averaged 70.07. In governance terms, Botswana scores particularly well, especially in combatting corruption." Botswana's performance is somewhat unusual and significant relative to its peers, especially when one considers its grim economic situation at independence, as detailed above by King (1966).

Botswana's political leadership has, until recently, been less inclined or interested in tapping into the fund, as the country had sufficient resources to advance its development agenda. Furthermore, there is no evidence of outright diversion of public resources by the political leadership to advance regional or sectional interests. However, there is an evident deficit in the regulatory framework of the Fund that made it open and potentially vulnerable to depletion. Because Parliament is not required to pass a budget that is consistent with the long-term objective of maintaining the Pula Fund at a certain level, it could approve a (legal) budget with a large deficit that depletes the Government's share of the Pula Fund. This would be legal in terms of public finance legislation, even though inimical to the goal of sustainable long-term saving for future generations.

It is a paradox that Botswana, one of the architects of the 2008 Santiago Principles that "promote transparency, good governance, accountability and prudent investment practices" (IFSWF 2019), does have some underlying issues with accountability, transparency, and good governance that bedevil the Fund. These require immediate attention to ensure a sustainable Fund that is in tune with its modest objectives. According to a former deputy governor of the Bank of Botswana, even though the Fund has not been excessively used, the process of appointing fund managers is not transparent (Interview, 26 Aug 2019). Botswana's lack of transparency has been overshadowed by the "country's history as a relatively efficient government with low levels of corruption," but it is a problem that makes independent citizens scrutiny of mineral revenues and spending, including those regarding the Pula Fund, more difficult (Jefferis 2016, p. 27). Jefferis comments further on this challenge:

> Botswana has not paid particular attention to the transparency of policy formulation and implementation. This may not have been a big issue when the underlying quality of policy was good and financial resources were abundant. However, when difficult decisions need to be made, transparency could arguably assist in building public understanding for some of the resulting choices. Transparency also contributes to the structure of checks and balances that underpin better decision-making. (Jefferis 2016, p. 34)

The Natural Resource Governance Institute (2013, p. 2) also contends that the Pula Fund is lacking in what it calls "good governance fundamentals such as transparency and having an effective oversight," among others. This suggests that although transparency of the Pula Fund is not yet a public or political issue, some legitimate concerns have been raised in certain quarters. According to a former governor of the Bank of Botswana the public has not shown interest in the management of the Pula Fund, and as such there has been no public participation, nor even questions raised about the Fund, as the public has confidence in the Bank of Botswana. Only a few politicians took an interest in it (Interview, 17 Sept 2019).

Issues of transparency regarding the Fund are not surprising, because this is not the only area in which Botswana faces a challenge of transparency. The opacity of the diamond contract between the state and De Beers also suggests, or supports, a transparency deficit in Botswana (Sharife 2016). Sharife (2016) further noted that it is not easy to determine whether Botswana's citizens have benefited as they should from the diamond wealth, as the mining contracts have been kept out of public view, declaring that the Botswana government, the Botswana Democratic Party (BDP), and De Beers

> have knit the political and corporate structures together in such a way that they undermine accountability and regulatory systems with a culture of secrecy (framed by De Beers as "confidentiality"). As a private entity, De Beers' dealings are largely protected from scrutiny. Unlike the EU and U.S., where governments once banned or prosecuted De Beers for price-fixing and other anti-competitive activities, Botswana's government and its ruling party have been direct collaborators.

In 2019, the opposition coalition, the Umbrella for Democratic Change (UDC), pledged to make the contracts accessible to the public if it were voted into power, a suggestion that the government publicly rejected. However, the government has maintained that Botswana has benefited most from the relationship with De Beers.

The other concern is that the Fund lacks a robust monetary target size, as it only saves what is left after the budget has been prepared. For instance, the accumulation in or withdrawal of monies from the Fund is driven by a combination of balance of payments and fiscal surpluses/deficits, and not by any explicit targeting of how large the Fund should be or any explicit allocation of a portion of mineral revenues to the Fund. The National Development Plan 11 proposes a new fiscal rule to be introduced in the medium term, that government should save 40% of mineral revenues for the future (Interview, 26 Aug 2019). However, this proposal does appear to have come rather late in the day. Saving 40% of mineral revenues could have been effective had it been introduced 30 years ago, but there is no commitment as to when this rule will be introduced (Interview, 26 Aug 2019). Intertwined with this, the Pula Fund is faced with an underlying challenge of preservation or protection, in part because it lacks sound or firm policy rules on inflows (deposits) and withdrawals, save for accounting procedural rules. A former deputy governor of the Bank of Botswana noted that "Although government has only access to its share of the Pula Fund as the other share is owned by the Bank of Botswana, there are no restrictions from withdrawing from the government share of the Pula Fund as long as the budget has been approved by parliament" (Interview, 26 Aug 2019). The Natural Resource Governance Institute (2013, p. 5) contends that "the Pula Fund is not governed by strict deposit and withdrawal rules. Instead, deposits are determined by the size of foreign exchange inflows and the size of fiscal surplus. Withdrawals cover fiscal deficits." Another respondent observed that there has been no need to define the rules as there has not been demand or pressure from politicians and that the country had sufficient resources to draw from domestically (Interview, 17 Sept 2019).

The absence of a specific monetary target size and defined rules regarding deposits and withdrawals explains in part why the Fund has been diminishing or declining over the years and thus vulnerable to being depleted. For instance, legitimate withdrawals were made from the government's share of the fund, causing a decline in the fund during the creation of the Botswana Public Officers Pension Fund (BPOPF) and during the global financial crisis. Ntibinyane (2016) has observed that some of the reasons put forward by the Bank and government for withdrawals are "financing the pensions of public servants, covering a budget deficit, boosting liquidity reserves and avoiding unpopular decisions such as increasing taxes, doing away with free education and freezing public servants' wages." In some instances, the reasons for withdrawals are not stated (Ntibinyane 2016). Ntibinyane further noted that "in 2002, a drawdown of P8 billion was made to float a government employees' pension fund. A further P5.2 billion was withdrawn the following year for undisclosed reasons. In 2005, an additional P11 billion was withdrawn" (roughly, US$1 = Pula 11.97). Ntibinyane has further pointed out that "P8.1 billion was withdrawn to compensate for the effects of the 2008 recession. Another drawdown of P2 billion was made in September 2010 to finance back pay for civil servants and, in 2012, R18.1 billion was withdrawn for what the ministry of finance said was for greater import cover. In 2014, a withdrawal of P1.8 billion was not explained." Thus, the point for concern is not the withdrawals per se, but the "unexplained" and "opportunistic" drawdowns (Ntibinyane 2016). Indeed, these types of withdrawals explain, in part, why the Fund has not performed as expected.

Moreover, the fund managers have no control over Botswana's export earnings from diamonds, or how much of those earnings are paid into the Pula Fund. Furthermore, the lack of clear rules creates a crisis, as the Pula Fund has been shrinking (Interview, 26 Aug 2019; also see Table 1). It should also be noted that since the Pula Fund is under the purview of the Bank of Botswana regarding its management, the absence of sound withdrawal rules and a lack of a monetary target size allows the government to make withdrawals with ease.

Despite the aforementioned weaknesses of the Pula Fund, its operation has not been characterized by open political conflict: just as Botswana's

Table 1 Pula Fund Assests, 1997–2018

As at year end	*Pula Fund*			*GDP Pmn*	*Pula Fund*	
	Total assets Pmn	*o/w GoB Pmn*	*GoB Share %*		*Total % of GDP*	*GoB share (%)*
1997	17,654	14,365	81.4	18,328	96.3	78.4
1998	25,562	18,233	71.3	20,244	126.3	90.1
1999	24,454	18,074	73.9	25,361	96.4	71.3
2000	28,712	23,081	80.4	29,531	97.2	78.2
2001	32,176	27,565	85.7	32,066	100.3	86.0
2002	24,473	15,553	63.6	34,416	71.1	45.2
2003	19,245	9,507	49.4	37,182	51.8	25.6
2004	20,013	8,893	44.4	42,037	47.6	21.2
2005	24,867	11,637	46.8	50,752	49.0	22.9
2006	36,855	19,967	54.2	59,107	62.4	33.8
2007	39,722	26,483	66.7	67,153	59.2	39.4
2008	51,626	30,456	59.0	74,721	69.1	40.8
2009	43,530	21,818	50.1	73,462	59.3	29.7
2010	44,733	13,552	30.3	86,867	51.5	15.6
2011	51,461	21,637	42.0	104,980	49.0	20.6
2012	39,353	18,231	46.3	109,870	35.8	16.6
2013	49,327	26,161	53.0	125,158	39.4	20.9
2014	54,758	37,261	68.0	145,869	37.5	25.5
2015	61,183	35,005	57.2	146,066	41.9	24.0
2016	54,146	28,692	53.0	170,564	31.7	16.8
2017	55,998	30,279	54.1	180,102	31.1	16.8
2018	47,470	23,991	50.5	189,869	25.0	12.6

Source Bank of Botswana Annual Financial Statements, various years; Statistics Botswana National Accounts

minerals have been peacefully extracted with little or no conflict, the domestic decision to establish the Pula Fund and the form it took have not been associated with any political conflict. Similarly, there has not been any conflict on the operation of the Fund regarding its objectives, when to deposit or withdraw from the fund or even the products in which to invest. This is the case, even though the creation of an SWF is a political decision with the potential to ignite conflict. Issues of control, power, and resistance to the creation of the Fund and its management have not arisen either. The critical question is why Botswana has experienced no political conflict, even if the fund is underperforming and fails to show good governance elements as discussed below. Furthermore, inequality in Botswana is the third highest in the world, demonstrating that not all citizens are benefiting from the diamond resources. This article attributes the absence of conflict to a number of factors, including institutional factors, which are discussed below.

The first such factor is the nature of the political elite and the type of political leadership Botswana has had and has sustained since its formative years. Botswana had a selfless political leadership that has, in the main, demonstrated its commitment to transforming the country, its economy, and the lives of its citizens through its policies that were meant to benefit everyone, as opposed to sectarian interests. Second, the absence of conflict and resistance could also be explained by the fact that minerals in Botswana are not only owned by the state but are used for the benefit of citizens rather than for sectarian purposes. The third factor is a lack of ambiguity regarding ownership of mineral revenues, as this is clearly defined: these revenues are owned by the government and not by resource communities or people. This has ensured that even areas that are not endowed equally benefit, irrespective of where the resources are found or extracted.

Other factors are institutional. The fourth factor in the absence of conflict in relation to the Pula Fund is the dominance and overbearing effect of the executive over the legislature, which has been widely documented in academic literature. Molomo (2000) stated that the extensive powers of the President extend to other branches of government. In addition to the overbearing effect of the powers of the executive, the legislature is controlled by the ruling party (Good 1999; de Jager and Sebudubudu 2017) that requires its members to uphold party caucus decisions. This curtails and frustrates debate in the legislature. In addition, members of cabinet are drawn from Parliament, and are more in terms of numbers than ordinary members of Parliament. Consequently, Good (1999) declared that non-accountability has become the standard behavior. It is, therefore, understandable and fitting that some scholars have described Botswana's legislature as "a mere registration chamber" of executive decisions, on account of its weaknesses and lack of independence (Sebudubudu et al. 2013, p. 33). In this sense, the institution that should take a leading role in discussing the Pula Fund has been mutated. The weakness of Botswana's legislature became more apparent during Ian Khama's presidency between 2008 and 2018, a period that saw regression in several indices due to fears of Khama's intolerant leadership style.

The other factor that accounts for the dearth of obvious conflict or opposition to the Pula Fund is the fragility of civil society. Given that the legislature has been reduced to "a mere 'registration chamber' of executive decisions" (Sebudubudu et al. 2013, p. 33), civil society, another critical force in a democracy, has not been spared from government control and monitoring. Control was more metered in those that are considered threats to the government, such as labor unions and private media (Mogalakwe and Sebudubudu 2006), because the state media unflaggingly supports the party line. Equally, Good (2003, p. 5) stated that civil society organizations were adversely perceived, and consequently "induced an apoliticism, a tendency towards self-censorship and a dependency posture within NGOs." Furthermore, Good (2003, p. 6) observed that the state did not welcome "the activity of civil society in the broad and vital areas of human rights, democracy, and political education and the San," Botswana's Indigenous minority people who were forcefully removed from their traditional lands to allow for diamond mining. With this uncertain attitude of the state in these critical areas, it is hard to imagine how civil society could grow, as Botswana's political culture eased government control. We argue that government policies have hindered the growth of civil society, which in part aided sustenance of the status quo. The Bertelsmann Transformation Index (2018) is apt and telling on why civil society has remained stagnant in Botswana, noting that it.

> has upheld its tradition of public consultation and consensus-building through traditional council meetings (Kgotla). However, the absence of a strong civil society is evident, owing in part to government policies that work against the development of strong civil society organizations (through legislative obstacles). The country's political culture also contributes to passive participation in matters of governance as it promotes submission to

> those in positions of leadership and authority. Those who hold alternative views are viewed with an element of suspicion and distrust by government; in most instances their views are not covered by state media as they are often associated with the opposition. These include among others trade unions, private media and even intellectuals.

In a small country such as Botswana, it is not difficult for the state to produce an inactive civil society because the state is the main source of resources. The consequences for those who oppose the state can be dire. For instance, a subtle advertising ban on privately owned media has been put in place, sending a clear and chilling message that in order to benefit from state resources, one must portray the state positively. This extends even to appointments in senior positions and awards of contracts or tenders. These measures became more pronounced under Ian Khama's government. The result of such actions is that critical voices of civil society, and the population as a whole, are subdued. Under these circumstances, the net effect is that the government side dominates the rest, including and especially those few alternative voices. The above descriptions of state–society relations are crucial to understanding the operation and governance of the Pula Fund. They help to deepen our understanding of the role of political culture in shaping political debates generally in Botswana, and an inference can be made that these play a central part in the lack of obvious conflict over the operation of the Pula Fund.

Overall, the weaknesses of civil society, subtle government censorship leading to some form of general self-censorship by the populace, civil society, and critical media, and the salient roles of traditional institutions have reduced the space of forces that work for the advancement of democracy. With these factors at play in the general body of politics, it is not surprising that the operation and governance of the Fund has generated little opposition or conflict, as the institutions that were supposed to take the lead were somewhat subdued.

In the absence of conflict, no crisis of legitimacy has arisen, so far, since the Pula Fund was created, as to whether to spend or save the funds. Equally, no evident fears of abuse or misappropriation of funds have become political or public issues, but frequent withdrawals have been noted. A former Governor of the Bank of Botswana observed that the Pula Fund has not been overly or excessively used because the country had adequate domestic resources to draw from for its needs (Interview, 17 Sept 2019). However, there has been pressure in certain quarters to withdraw more money from the Fund. For instance, opposition parties have consistently and frequently argued that the Pula Fund monies should be used to finance development projects (Interview, 17 Sept 2019). The government has, to some extent, rebuffed these suggestions. However, it is important to note that the Pula Fund has not been used to advance sectarian interests (Interview, 17 Sept 2019), in part because Botswana has relatively functional institutions. For instance, its central bank, the Bank of Botswana, is one of the best managed in Africa because the Act guarantees its independence, and the government adheres to the law. Consequently, the Fund's management has not been a subject of public controversy, even though it could be argued that the Bank has, in the main, escaped public scrutiny due to the difficulty of obtaining information about it, except what is authorized by the Board as a result of the oath of secrecy that prohibits members, officers, or employees of the Bank, or the Bank's appointed auditors, from disclosing information to any person (Republic of Botswana 1996). Nevertheless, it should be pointed out that the Bank of Botswana has been led by an experienced and respected governor, Linah Mohohlo, who was succeeded upon retirement by one of the deputies, Moses Pelalelo, in October 2016, ensuring continuity and stability in the Bank's leadership.

It is important to note that even though it has not generated open conflict, the operational challenges of the Fund have put its sustainability under suspicion, particularly the realization of its objective as a future generation's fund. A former Deputy Governor of the Bank of Botswana argues that the Pula Fund is only sustainable to the extent that it functions as a stabilization but not as an intergenerational fund, as it has not been growing in keeping with the size of the economy. According to Table 1, the Pula Fund and the country's Gross Domestic Product (GDP) stood at 47 billion pula and 189 billion pula in December 2018, respectively, which is in contrast to the P25 billion and P20 billion of the Pula Fund and GDP in 1998, respectively. Table 1 further shows that the total value of Pula Fund assets grew steadily from 17 billion pula in 1997 to P61 billion in 2015 before declining to P47 billion in December 2018. The government share of the fund was P14 billion (81%) in 1997 and P9 billion (44%) in 2004; it was highest at P37 billion (68%) in 2014 and declined to P23 billion (51%) in 2018. The Pula Fund has also not grown in line with the size of the economy, that is, as a percentage of GDP. For instance, it was 96% of the GDP in 1997, 48% in 2004, 42% in 2015, and 25% in 2018. The government share as a percentage of GDP was put at

78%, 21%, 24%, and 13% in 1997, 2004, 2015, and 2018, respectively. Withdrawals were made from the Fund, especially in recent years to stabilize the economy, which is not necessarily a bad thing as such short-term stabilization of the economy is one of the unstated role that the Fund is expected to perform (IFSWF 2019). The Fund's lack of growth could be for a number of reasons beyond withdrawals. This chapter thus contends that Botswana lacks a deliberate strategy and commitment to grow its Pula Fund in the same way that countries such as Norway did.

Despite being the oldest SWF in Africa, the Pula Fund remains modest because of the challenges confronting it, including the lack of a deliberate strategy and monetary target size for the fund to grow in the same way as other funds such as that of Norway. The Fund has been criticized for its "lackluster performance" which is in part because of several withdrawals made by the Bank and the Ministry of Finance (Ntibinyane 2016). In terms of Table 1, the value of the Pula Fund was around 47 billion pula as of December 2018, and shares for the Bank of Botswana and the government stood at almost 50% each. Ramokopelwa (2019) noted that the fund had a value of 55 billion pula the previous year, and declined to P49.9 billion in February 2019. Overall, Table 1 shows that the Pula Fund has been declining in total value and has not been growing in relation to the size of the economy. This might be due to the lack of clarity about objectives, the lack of deliberate strategy and monetary target size, and the lack of rules regarding withdrawals and deposits, but it may also be due to the lack of sound governance or leadership.

The political and economic outcome has been a relatively beneficial, consistent, and stable fund; and as such, there has been little if any opposition to or conflict with the Fund. Instead, the Fund has received ongoing support from the political leadership since its creation because of its envisaged and consequent benefits (Interview, 17 Sept 2019). The absence of opposition to the Fund could be explained by the fact that the Fund did not take away many resources required for other pressing needs of Botswana's development, as it takes what remains after the budget has been allocated.

According to a deputy governor and a former governor of the Bank of Botswana, the Fund was created at a time of financial surpluses, and those surpluses were accumulated into foreign reserves. As such, the then-Ministry of Finance and Development Planning and the Bank of Botswana made the joint decision to manage funds as an investment (Interview 26 Aug 2019; 17 Sep 2019). From the beginning, Botswana has maintained a conservative stance toward its sovereign wealth fund without major changes to the Fund's original objectives and the way it has been managed. Although the fund has not changed its objectives on paper, its operation is different from those objectives. In May 2019, the Bank of Botswana expanded its mandate to go beyond a small number of foreign exchanges, given the changing global economy. However, the Fund faces some challenges, as noted above. It also appears to have partially met its (unstated/undeclared) objective of acting as a stabilization fund. However, its sustainability is doubtful as a future generation's fund. There has been little, if any, pressure for change, and reforms of the fund have been minimal, as the expansion of its mandate in May 2019 has demonstrated, because the same governing regime that created the Pula Fund is still in power. This suggests that the Fund is likely to continue for the foreseeable future. Even so, critical voices have been raised on deficits confronting the Fund, in certain quarters.

Despite its widely recognized and unusual achievements, as demonstrated by its social indicators, and its reputation for good governance, Botswana has not been successful in weaning itself from resource dependency and a narrow economic base, so that it is not immune to external shocks. The most recent of such shocks, with far-reaching effects and implications for the economy and for Botswana as a whole, were experienced during the 2008 financial crisis, which put more pressure on the country to move away from a limited industry/economic base. Such a shift has not been easy for Botswana, despite concerted efforts in that direction.

Conclusion

Botswana has maintained its long-term investment fund, the Pula Fund, since its establishment in the early 1990s. Botswana's Pula Fund has not faced any internal political opposition, nor has it been a source of conflict as it received political support. Botswana has sustained the Fund as a modest fund with unchanged objectives, as it would appear there was no pressure to change them, despite a deteriorating performance record and critiques of its rules and governance structures. However, there is no doubt that the Fund has served as a cushion during difficult periods.

Overall, the policy framework within which the Fund operates has been found wanting in a number of areas. The Pula Fund is considered weak as it lacks sound oversight mechanisms, clearly defined rules on inflows and withdrawals, and open operations. All these weaknesses make it vulnerable to possible depletion, and may have contributed to its recent decline in size, as there is no defined monetary target size because it saves what remains after preparing a budget. Furthermore, all these weaknesses or deficits could be functions of a country that became a victim of its own success. Amid sufficient resources resulting from a mineral boom, Botswana may have been complacent in the management of its SWF and in the process failed to introduce reforms that could have enhanced its performance. The other challenge facing the country is its continued dependence on a narrow economic base despite concerted efforts to enhance the country's economic base from minerals.

In conclusion, it is striking that the Pula Fund has been successful in terms of suppressing controversy and political conflict, but unsuccessful in terms of its financial performance, if measured by its potential value. This might be due to the lack of clarity over objectives, monetary target size, or rules regarding withdrawals and deposits, but it may also be due to the lack of sound governance or leadership. Given that Botswana is touted as a leader of good governance and leadership in Africa, its success has not transcended to the fund's success/fiscal performance. However, the fund's success should be understood in the context of its objectives.

References

Ahmadov I, Tsani S, Aslanli K (2011) Sovereign wealth funds as the emerging players in the global financial arena: Characteristics, risks and governance. In: Sovereign wealth funds: New challenges for the Caspian countries. Revenue Watch Institute, Baku

Bank of Botswana (2019) Pula Fund. http://www.bankofbotswana.bw/content/2009103013033-pula-fund. Accessed 2 Sept 2019

Bertelsmann Transformation Index (BTI) (2018) Botswana country report. https://www.bti-project.org/en/reports/country-reports/detail/itc/bwa/ity/2018/itr/esa/. Accessed 23 Jan 2020

Coutinho L (2011) The resource curse and fiscal policy. Cyprus Economic Policy Review 5(1):43–70

De Jager N, Sebudubudu D (2017) Towards understanding Botswana and South Africa's ambivalence to liberal democracy. Journal of Contemporary African Studies 35(1):15–33. http://www.tandfonline.com/doi/full/10.1080/02589001.2016.1246682

Good K (1999) Enduring elite democracy in Botswana. In: Burnell P, Calvert P (eds) The resilience of democracy: Persistent practice, durable idea. Frank Cass, London, p 50–66

Good K (2003) Bushman and diamonds: (Un)civil society in Botswana. Discussion Paper 23, Uppsala

Government of Botswana (1996) Bank of Botswana Act. Government Printer, Gaborone

Harvey R (2015) What role for natural resources in Botswana's quest for economic diversification? SAIIA Policy Briefing 135

Humphreys M, Sandbu ME (2007) The political economy of natural resource funds. In: Humphreys M, Sachs JD (eds) Escaping the resource curse. Columbia UP, New York, p 194–233

IFSWF (2019) Santiago principles. https://www.ifswf.org/santiago-principles-landing/santiago-principles

Interview with a former deputy governor of Bank of Botswana, 26 Aug 2019

Interview with a former governor of Bank of Botswana, 17 Sept 2019 & 3 Oct 2019

Jefferis K (2016) Public finance and mineral revenues in Botswana. Technical Report; Government of Botswana and World Bank: Econsult Botswana (PTY) LTD

King C (1966) Bechuanaland poor, hungry desert land, independence to come Friday. Southern Africa News Service. Botswana as it was seen 45 years ago. https://www.letakasafaris.com/botswana-as-it-was-seen-45-years-ago/2012/. Accessed 18 Oct 2019

Leith JC (2005) Why Botswana prospered. McGill-Queen's UP, Montreal

Lewin M (2011) Botswana's success: Good governance, good policies, and good luck. In: Chuhan-Pole P, Angwafo M (eds) Yes Africa can: Success stories from a dynamic continent. The World Bank, Washington, DC, p 81–90

Meijia XP, Castel V (2012) Could oil shine like diamonds? How Botswana avoided the resource curse and its implications for a new Libya. African Development Bank (AfDB)

Mogalakwe M, Sebudubudu D (2006) Trends in state-civil society relations in Botswana. Journal of African Elections 5(2):207–224

Molomo MG (2000) Democracy under siege: The presidency and executive powers in Botswana. Pula: Botswana Journal of African Studies 14(1):95–108

Natural Resource Governance Institute (2013) Natural Resource Funds: Botswana Pula Fund. Columbia Center on Sustainable Investment. https://resourcegovernance.org/sites/default/files/NRF_Botswana_July2013.pdf. Accessed 3 Sept 2019

Natural Resource Governance Institute (2014) Managing the public trust: How to make resource funds work for citizens. Natural Resource Governance Institute and Columbia Center on Sustainable Investment

Ntibinyane N (2016) Botswana repeatedly raids preservation fund. Mail & Guardian. https://mg.co.za/article/2016-02-04-botswana-repeatedly-raids-preservation-fund/. Accessed 23 Jan 2020

OECD (2013) OECD investment policy reviews: Botswana. https://www.oecd.org/daf/inv/investment-policy/IPR_Botswana_Oct2013-Summary.pdf. Accessed 23 Jan 2020

Ramokopelwa K (2019) Sovereign wealth funds must be closely guarded. Botswana Guardian, 21 May. http://www.botswanaguardian.co.bw/news/item/4152-sovereign-wealth-funds-must-be-closely-guarded.html. Accessed 18 & 27 Oct 2019

Republic of Botswana (1996) Bank of Botswana Act. Chapter 55:01

Rotberg RI (2003). The roots of Africa's leadership deficit. Compass: A Journal of Leadership 1(1):28–32

Sebudubudu D (2019) Botswana. In: Adetula V, Kamsi B, Mehler A, Melber H (eds) Africa yearbook: Politics, economy and society south of the Sahara in 2018, Volume 15. Brill, Leiden and Boston, p 436–444

Sebudubudu D, Bodilenyane K, Kwerepe P (2016) The politics of opposition electoral coalitions and alliances in Botswana. African Review 43(1):1–25

Sebudubudu D, Botlhomilwe MZ (2012) The critical role of leadership in Botswana's development: What lessons? Leadership 8(1):29–45

Sebudubudu D, Maripe B, Botlhomilwe MZ, Malila IS (2013) The mutation of Parliament into a "registration chamber": Executive dominance over the legislature in Botswana. African Review 40(2):33–59

Sebudubudu D, Mooketsane K (2016) Why Botswana is a deviant case to the natural resource curse. African Review 43(2):83–95

Sebudubudu D, Molutsi P (2011) The elite as a critical factor in national development: The case of Botswana. Nordic Africa Institute, Uppsala

Sharife K (2016) Botswana's diamond deception. The Open Society Initiative for Southern Africa (OSISA), Southern Africa Resource Watch (SARW) and the African CIR. https://s3.amazonaws.com/rgi-ocuments/8c176b2c63140d25c086b315c2455f5ba07b204d.pdf. Accessed 23 Jan 2020

Transparency International (2019) Botswana. https://www.transparency.org/country/BWA. Accessed 16 Dec 2019

Wills SE, Senbet L, Simbanegavi W (2016) Sovereign wealth funds and natural resource management in Africa. Journal of African Economies 25(2):ii3–ii19

World Bank (2019) The World Bank in Botswana. https://www.worldbank.org/en/country/botswana/overview. Accessed 15 Nov 2019

Chile: A Successful Story in Latin America?

Xu Yi-chong and Diego Leiva

Introduction

By the time the center-left coalition finished its two-decade rule in 2010, Chile had become the envy of South America as it made the transition, planned and implemented by the military regime in 1990, progressed steadily to a stable democracy. One key contributor to Chile's stable transition was the change of its economy from an "authoritarian liberal exclusive model" to a "democratic liberal inclusive model" (Angell 1993). After the return to democracy in 1990, the government continued many components of the economic development model based on liberalization, privatization, and deregulation, and a tight fiscal discipline, while increased social spending was accompanied by tax increases. The average annual GDP growth of 2.4% during the 16-year Pinochet regime rose

The original version of this chapter was revised: This chapter source of Table 1 has been updated. The correction to this chapter is available at https://doi.org/10.1007/978-3-030-78251-1_12

Xu YC (✉) · D. Leiva
Griffith University, Nathan, QLD, Australia
e-mail: yi-chong.xu@griffith.edu.au

E. Okpanachi and R. Chowdhari Tremblay (eds.), *The Political Economy of Natural Resource Funds*, International Political Economy Series, https://doi.org/10.1007/978-3-030-78251-1_5

to 5.8% on average in 1990–2007; inflation dropped from an average of 27% under the dictatorship to 3% by 2005, and remained at that level; real wages stagnated during the dictatorship but doubled in the following decade; and the unemployment rate had fallen to 5%. In 2010, Chile became a member of OECD, and was considered by the IMF and the World Bank as a model for the economies in the region and for resource-based economies around the world.

The cornerstone of Chile's macroeconomic growth and fiscal stability was its structural balance rule that allowed the government to build up its fiscal capacity to manage the cycle of commodity prices and buffer it against external shocks, including the global financial crisis (GFC) in 2008–2009. Its two natural resource funds (NRFs)—the Pension Reserve Fund and the Economic and Social Stabilization Fund (ESSF) built on the copper Stabilization Fund of the late 1980s—played key roles in helping the government fend off spending pressures when copper prices surged, and in helping stabilize the fiscal cycle in the face of large and unexpected falls in government revenues caused by copper price volatility and global shocks. After the crisis, Chile continued to allocate fiscal revenues to the NRFs for future negative economic cycles, in contrast to other economies in the region whose funds were used without being replenished or ceased to exist completely. In line with other country studies in this collection, this chapter seeks to answer three questions: why the two NRFs were created in Chile; how they were governed; and how they operate. To contribute to the literature on the Chilean NRFs (Ffrench-Davis 2016; Parrado 2017; Saffirio 2013), this chapter goes beyond Chile's oft-cited "success" story and examines the debates over the usage of the funds in 2009, 2014, and 2016. The finding suggests that, however insulated, decisions on whether to withdraw resources from NRFs tend to be politically and technically controversial. They are politically contentious because these funds are owned and managed by the state. They are technically controversial because those managing the fund tend to see it as their responsibility to protect and expand it while those desiring to use it have much broader policy agendas in mind. This is not unique to Chile. The debates over whether and when to withdraw resources from NRFs, and how much to withdraw, are nonetheless shaped by the politics of the day; as there is no unified form of NRFs, there is also no single way of managing them.

This chapter begins with a brief summary of the Chilean economy over the last 40 years, its main characteristics and phases, followed by the establishment of the Pension Reserve Fund and the Economic and Social Stabilization Fund in 2006 as part of broader economic reforms.

The second section explains the governing structure of the two funds within Chile's rule-based fiscal framework. The third section discusses the operation of the two funds, and specifically explains the debates over and different outcomes of the use of the funds to address fiscal imbalances in 2009, 2014, and 2016. The chapter concludes with an assessment of the Chilean case as a "successful story" among NRFs in Latin America.

The Chilean Economy: An Overview of the Past 40 Years

According to Ocampo (2014), "Financial crises, in their different dimensions—external debt, balance of payments and banking crises or, frequently, a mix of all of these—have been a recurrent phenomenon in Latin America's economic history." During the Latin American debt crisis in the 1980s, the Pinochet government adopted a series of economic reforms designed and implemented by a group of economists known as the Chicago Boys (Undurraga 2015). Drawing lessons from the state-led industrialization in the 1950s and 1960s and the hyperinflation of the 1970s, these US-educated and trained economists concluded that the role of the state had to be reduced. Their economic reform measures included privatization, deregulation, elimination of trade barriers in favor of free trade, opening its financial sector for domestic and international competition, limitations on trade union activities, and imposition of strict fiscal rules to control inflation (Brender 2010; Fourcade-Gourinchas and Babb 2002). The implementation of these reform policies, which were later known as the Washington consensus, by Pinochet's dictatorship helped control inflation in Chile (Ffrench-Davis 2003). Its economic development in general, however, suffered, with an average GDP growth of 2.9% and rising inequality due to increase in wealth concentration and cuts in social spending. Meanwhile, the Chilean economy remained vulnerable to external shocks because of its heavy reliance on copper exports and with little investment made to diversify its economy.

Chile's return to democracy in 1990 did not fundamentally alter its economic structure. The "Concertación" governments of the "center-left" coalition in the following two decades, covering the governments of Patricio Aylwin (1990–1994), Eduardo Frei (1994–2000), Ricardo Lagos (2000–2006), and Michelle Bachelet (2006–2010), adopted the approach of "change with continuity," maintaining the market-oriented economy by keeping it open for trade and investment, relying on private sector competition, and restricting the role of government in economy (Fernandez and Vera 2012; Siavelis 2009; Undurraga 2015). Most importantly, the successive elected governments continued a strong commitment to sound economic management and fiscal responsibility. At the time, the political situation in Chile was unstable and the sustainability of democracy was by no means assured. Instead of relying on spending to meet the public demand, the democratic center-left governments emphasized responsible economic policies with strict fiscal disciplines. This was due in part to the lesson from its own history (one of the main conditions of the pretexts of the military's takeover in 1973), and in part to the harsh consequences other Latin American countries faced from the debt crisis in the 1980s. A small group of high-caliber economists dominating decision-making made the government's commitment to tight fiscal policies possible.

By the early 2000s, Chile had the highest per capita income growth in Latin America; inflation fell from 30% at the beginning of the 1990s to 3% between 1999 and 2002; and government revenue reached 3% of GDP. The fiscal surpluses averaged 1.7% and 1.2% of GDP in 1990–1993 and 1994–1999. The national government-debt-to-GDP ratio was down from 45% in 1990 to 15% in 2001. Meanwhile, the center-left coalition governments adopted a series of reforms, including tax reform that increased corporate tax and a new labor law that restored several workers' rights, raised minimum wages, and brought workers and businesses to the negotiation table (Ffrench-Davis 2003). With the tax reform, the government was able to shift toward a massive redistribution of wealth in housing, health, education, and poverty reduction in the form of government transfers (Engel et al. 1999). By 2006, for instance, primary education had become universal; secondary education had reached 90%, and higher education had doubled from its level in 1990. Living standards improved across the board: the poverty rate was down from 45% in 1987 to 21% in 2000; 72% of Chileans had become homeowners; infant mortality went down to 10 per 1,000 and life expectancy reached 80 years for women and 71 years for men. These reforms in redistribution policies and stable macroeconomic and fiscal policies contributed to the improvement of the functioning of markets and enabled its sustained economic growth. Chile was considered a potential candidate to join the OECD.

The success of economic and social developmentin Chile in the 1990s did not change its heavy reliance on copper production and exports. Chile is the world's largest copper producer, with an annual production of 5.5 million metric tons, which accounted for over 40% of world production. Copper accounted for about 40–45% of its exports. As copper production was one of the most important sources of revenues for the

Chilean government, the volatile income posed serious challenges for fiscal stability, as seen during the Asian financial crisis. Chile experienced large net capital inflows (more than 10% of GDP in 1997), but its trade balance started to deteriorate after 1997 (Aggarwal 2000). The annual GDP growth rate declined from 6.6% in 1997 to −0.8% in 1999; unemployment rose from 6% to over 10%; and fiscal balance went to negative. In combination with a significant fall in copper prices and the induced terms of trade shock, the current account deficit reached over 5% of GDP in 1998 (DeShazo 2005; OECD 2003, 2010, 2012). Despite the external shocks and their effects, the government's commitment to fiscal disciplines and macroeconomic stability was unswayed, allowing the government to gradually reduce the level of national debts.

In 2001, to show its political commitment, the Chilean government adopted the structural balance rule under which it would operate with a structural budget surplus of 1% of GDP each year. The idea behind this structural balance rule was simple and non-controversial: when the economy performs above its projected long-term potential, the government should return a nominal surplus; and when the economy performs below the potential, the government could return a nominal deficit. Over the cycle, these should balance each other out (Lienert 2010). The structural balance rule was particularly important for an economy that relies on commodity production and exports, and its implementation would allow the government to insulate public spending from short-term copper price fluctuations and cyclical fluctuations of the economic activity (Chilean Ministry of Finance 2014a).

Other factors behind Chile's economic success in the 1990s and early 2000s included an independent and efficient central bank, stable management of the macroeconomic agenda, a sustained commitment to free trade and economic openness, an institutional environment committed to promoting investment, a consensus on tax policy, a dynamic private sector, and above all, a regime that kept an overall level tax above the average in the region, a broadly shared commitment to social spending directed at reducing poverty, especially in the poorest sectors, and transparency and honesty in government. Despite its success, Chile suffered from persistent rates of income inequality and distribution, due to a large extent to the quality of public education (OECD 2003).

Chile's NRFs: Process of Establishment, Characteristics, and Performance

In September 2006, the Chilean government upgraded the structural balance rule of 2001 and enacted the Fiscal Responsibility Law (FRL). The earlier rule had helped preserve fiscal discipline, ensure an annual fiscal spending level consistent with the central government's structural revenue, maintain a counter-cyclical fiscal stance in an environment of rising copper prices, and stabilize the level of public debts. Externally, it increased the Chilean government's credibility as an issuer of international debt, reduced the sovereign premium it had to pay, improved access to foreign financing in the face of negative external shocks, and minimized contagion from international crises. Domestically, it ensured the financial sustainability of public policies and facilitated long-term planning. Meanwhile, several issues indicated the need to refine the structural balance rule and led to the enactment of the FRL in 2006.

Among these issues were the perennial challenges governments face in calculating and forecasting copper prices and revenues and assessing the real fiscal risk factors that affect its budget decisions. In an economy in which expenditures were considered residual and national revenues heavily depended on copper production and exports, calculations of the structural balance were performed by adjusting the revenue side of the budget. It was a sensitive exercise. At the time these calculations were done in-house by the Ministry of Finance. The opposition parties and other commentators voiced skepticism about the procedures and their outcome, but they were presented in such a way that made it so difficult for non-specialists to understand (Blöndal and Curran 2004). The government wanted to change this impression to further demonstrate its commitment to fiscal discipline.

Another concern was the related issue of managing the unpredictable copper price and thus tax revenues. In 2005, the government introduced a small profit-related royalty as part of Chile's tax regime, in addition to the standard corporate income tax, which was levied on all mining companies, including the Chilean state-owned copper mining company CODELCO. This change in the tax regime transferred CODELCO to the central government budget. When copper prices surged in 2005–2006 to five times of that in 2001, even with the structural surplus balance rule (1% of GDP) copper revenue would have put pressure on absorptive fiscal capacity and triggered pressures for a sizable real appreciation of the currency, if fully spent in real time. These were the long- and short-term concerns behind the enactment of the Fiscal Responsibility Law in September 2006 (Chilean Ministry of Finance 2008).

The adoption of the FRL was highly praised by both OECD and the IMF as an important step to strengthen Chile's rule-based governing and institutionalization of key aspects of the structural balance and fiscal policy framework. It demonstrated the political commitment that the government would be "responsible before Parliament and the public for providing clear information on past fiscal outcomes, expected fiscal developments, and the policies proposed to deal with any fiscal imbalances and deviations from an agreed medium-term fiscal strategy" (Lienert 2010). The Constitution lays out the requirements for legislative authorization of spending and taxation, as well as the roles and responsibilities of the Parliament and the President as the executive in the budget process. The FRL 2006 further details the budgeting process: within 90 days of taking office, the President must adopt a decree defining the fiscal fundamentals of his/her administration and the expected effects of those fundamentals on the structural balance. Congress cannot increase or reduce the revenue estimated in the draft budget by the president, but can only reduce draft expenditure, so long as it is not permanent (Blöndal and Curran 2004; Rodriguez et al. 2007). This hierarchical structure, which limits decision-making on fiscal policy to a small group of technical experts, has been the hallmark of budgeting in Chile (Vammalle and Rivadeneira 2017), and has made it easier for Chile to meet the political commitment of fiscal disciplines, as compared to many other countries.

One key component of the FRL 2006 is the creation of the Fund for Economic and Social Stabilisation (Fondo de Estabilisación Económica y Social), or ESSF and the Pensions Reserve Fund (Fondo de Reserva de Pensiones, or PRF), both of which provided more durable foundations for maintaining the fiscal structural balance. Neither the PRF nor the ESSF was new: the PRF dates back to 1981, when the Chilean government introduced a mandatory pension plan under which individuals were required to deposit in pension funds an amount equal to 10% of their wages, and to make an additional contribution of 2–3% of their wages as a premium for disability and term life insurance as well as to cover administrative costs. Under this scheme, employers made no contributions. In 2002, the government introduced a system of individual unemployment insurance into which both employers and employees had to pay. Other related pension reforms were adopted too, such as basic old-age pension, low-income pension, and disability insurance income (Chilean Ministry of Finance 2009; De Mesa and Mesa-Lago 2006). Some of these schemes carried with them extensive contingent liabilities for the government. For example, the government guaranteed a minimum rate of return for the individual capital accounts—in effect, a state guarantee for all the firms that managed those accounts (Iglesias-Palau 2009).

The new PRF was meant to address these issues, especially to meet the government's "future" (after 2016) minimum pension commitment and social welfare spending. The PRF was established with an initial contribution of $604.5 million and would receive capital increase of an amount equivalent to 0.2% of the previous year's GDP. If the actual fiscal surplus exceeded 0.2% of GDP, the PRF would receive up to 0.5% of GDP revenue surplus through the Social and Economic Stabilisation Fund (ESSF). The PRF is a savings fund that is managed by the pension fund administration in order to ensure steady growth of assets and meet the government's pension liability (Valero and Monterde 2011).

The ESSF was preceded by the Chilean Copper Fund (CCF), sometimes known as the Copper Stabilization Fund, a fiscal stabilization fund created in 1985. This makes the ESSF different from many natural resource funds (NRFs) in several ways. First, unlike the natural resource funds in Alaska, Alberta, and Timor-Leste, which had been preceded by public campaigns and debates, the ESSF had no such campaigns because the CCF had been in operation for nearly two decades and Chile's elite decision-making in fiscal policy insulated the decision from the public. Such a strong top-down budgetary process led by the Budget Office of the Ministry of Finance, a small group of elite professionals, is a distinctive feature of budgeting in Chile, which is known for its rule-based and transparent governing and its long history of centralized fiscal policymaking (Schmidt-Hebbel 2012; Vammalle and Rivadeneira 2017).

Second, most NRFs were set up as savings funds to enable governments to provide future expenditures. This is often done for three reasons: intergenerational equity—that is, to provide income for future generations who should benefit from the country's natural resources that can and may become depleted in the near future; to limit current government expenditures to address the problems of inadequate absorption capacity; and to take precautions against large and unexpected shocks to the domestic or world economy. Savings funds, such as the NRFs in Alaska, Alberta, and Kuwait, are sometimes set up for all three reasons (Alsweilem et al. 2015; Bauer, ed. 2014). In Chile, the PRF was set up as a long-term saving fund to meet future pension and social welfare liabilities, whereas the ESSF was created as a stabilization fund to mitigate the volatility of the annual fiscal process. This is primarily because the volatility of world copper prices and uncertainties in copper production and exports presented serious challenges of public financial fluctuations. Other NRFs, such as that in Norway, may be established with the combined objectives of savings and stability, and such a distinction is important when considering the management of NRFs.

Third, the ESSF does not have a separate independent legal status, as do those in Alaska, Kuwait, and other sovereign wealth funds that have other sources of assets. The Ministry of Finance owns both the PRF and the ESSF and determines their investment and operational policies. The Central Bank of Chile (CBC) is the operational investment manager of both funds (including the allocation and oversight of external investment mandates in equities and corporate bonds), implementing a largely passive, index-driven investment policy (Alsweilem et al. 2015). This practice is considered inevitable in Chile because "the payments into and out of the fund are determined by decisions of parliament, often through the budgetary process, which is an integral part of the functions of the ministry of finance" (Bacon and Tordo 2006). The legal status of an NRF is not as important as the rules that are clearly articulated when the fund is established. A rule-based approach can avoid future political manipulation or even abuse of the fund; it also provides certainty for depositing additional resources to and withdrawing them from the fund, and thus provides predictability for fiscal policymaking (Fig. 1).

One key rule governing the assets and withdrawal is the principle of transfers into and out of the funds. As the ESSF serves as a tool to smooth out global and domestic business cycles, and government revenue is subject to volatility due to fluctuations of copper production and prices, forecasting prices is an important exercise. The ESSF continued and upgraded many of the CCF's practices, including its mechanism for forecasting the reference price. That is, when the world copper price was above a reference price, the additional revenues from copper exports by the state-owned copper company Corporación Nacional del Cobre (CODELCO) would be deposited into the CCF, which was managed by the Ministry of Finance. The reference price was determined by the Ministry of Finance based on the price average over the previous six years and "the trigger mechanisms for payments into and out of the fund were determined in agreement with the World Bank" (Bacon and Tordo 2006).

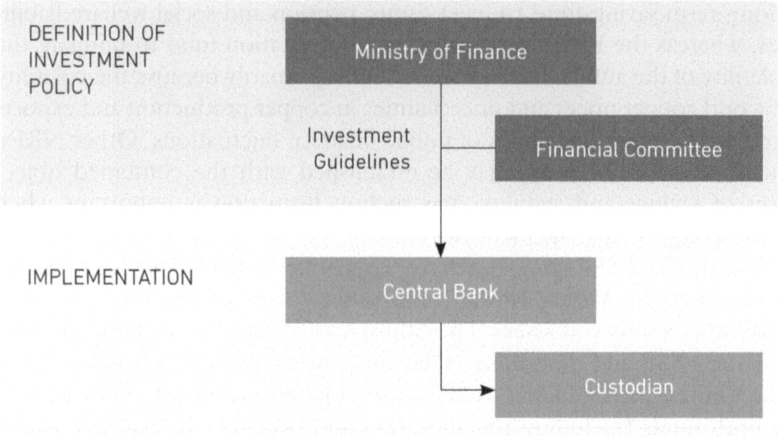

Fig. 1 Institutional framework of the Chilean SWFs (*Source* Chilean Ministry of Finance 2010)

This practice was criticized on two grounds. First, the long-term forecast price was entirely backward looking, with little consideration of a sharp drop in prices in the near future. Second, and more importantly, there was little public trust in the Ministry of Finance, and its calculations were too complicated for the public to understand. In addition, the oversight committee did not publish its report, and the central bank that managed the CCF did not issue statements on its performance, which further reduced the public accountability and trust on the fund.

To address these issues, in 2002 the government created committees of experts to provide independent technical estimates of potential copper output and the long-term copper price. Forecasting exercises were delegated to the copper price panel, which consisted of 14 individuals appointed by the minister of finance for one year at a time. The panel included a balance among members from the governing and opposition parties, employees of mining companies, and financial analysts of the sector. The rule for price referencing was also revised from backward looking to forward looking. Each panel member estimated the long-term (10-year average, instead of 6-year average as in 1985), forward-looking reference price of copper. At one of the two panel meetings held during the budget season, members submitted a forecast that was published anonymously. The two most extreme forecasts were discarded, and a simple average of the remaining 12 was used as a reference price without attempting to reach a consensus view. Based on the reference price, experts calculated revenues from potential output. The government then set a target for the structural balance. The price reference is a part and parcel of the responsibility of the ministry of finance over the PRF and the ESSF. Based on the price referencing provided by the experts, the Ministry decides the national budget and all major issues on investment and management of the PRF and the ESSF, including establishing investment guidelines, specifying the strategic portfolio allocation, defining the benchmarks for the funds' performance, and setting risk limits.

From an organizational perspective, the Ministry of Finance has the authority to decide whether the operational management of the ESSF will be carried out by the Central Bank of Chile, the Treasury (part of the Ministry of Finance), or other external fund managers. For instance, the CBC can serve as a fiscal agent of the Ministry of Finance and can manage the portfolio of the ESSF with the investment guidelines established by the Ministry. It can also delegate part of the fund's investment portfolio to external fund managers under the established investment guidelines and under the CBC's supervision. This practice is in line with those of many sovereign wealth funds, such as Australia's Future Fund and the Korea Investment Corporation (Cumming et al. 2017).

In 2007, the Ministry of Finance created an external advisory body, the Financial Committee, as part of the governing structure for the two investment funds. The Financial Committee consists of six independent macroeconomic and financial experts, appointed by the Ministry of Finance for an overlapping tenure of two years. The Committee meets every six weeks at the Ministry of Finance, which provides the secretarial work, including preparing technical reports on international financial conditions and performances of the PRF and the ESSF. It reviews the international and domestic financial developments and their implications for the performance of the two funds, evaluates fund managements, and, more importantly, advises the ministry of finance on the long-term investment policy of the PRF and the ESSF, and on other issues related to the two funds.

In sum, the PRF and the ESSF were set up in Chile in 2006 as part of broader economic reforms, supporting its sustainable fiscal and macroeconomic stability, and as an exercise of institutionalization of rule-based governing in Chile. The constitutional reform introduced in 2005 curbed military power, strengthened democratic institutions, and reduced the president's term from six to four years. The shorter presidential terms exposed the fiscal policymaking process to potential challenges and debates in Parliament, and therefore made it more accountable to the public. Expanding and strengthening the role and responsibility of elite professionals further cemented a relatively exclusive fiscal policymaking process in Chile that was not overly politicized, as was the case in its neighboring countries. In so doing, it also offered a greater degree of credibility and transparency.

Operation of the ESSF

The Fiscal Responsibility Law of 2006 clearly articulates and differentiates the mandates and investment policies of the PRF and ESSF and their relations, and provides the rules governing sources of funding for, deposits into, and withdrawals from the PRF and the ESSF. According to the FRL 2006, a minimum of 0.2% of the country's previous year's GDP must be deposited into the PRF annually. If the effective fiscal surplus exceeds this amount, the deposit amount can rise to a maximum of 0.5% of the previous year's GDP. The Ministry of Finance can also decide to deposit additional funds to the PRF. The ESSF is funded by the remaining amount of the effective fiscal surplus, which is the fiscal surplus after subtracting the contribution to the PRF, any funds used for public debt repayments, and estimated advanced payments into the ESSF (if any) in the previous fiscal year that had not materialized yet. The resources from the ESSF can be used at any time to finance authorized public expenditures when there is a fiscal deficit. Thus, by comparison, the ESSF has much more relaxed withdrawal rules than other NRFs, such as that of Alaska, whose rules require that the government deposit 25% of all mineral lease rentals, royalties, royalty sales proceeds, federal mineral-sharing proceeds, and bonuses received by the state into the Alaska Permanent Fund. The principal of the fund would be invested permanently and could not be spent without a further vote of the people, while a portion of the fund's income could be distributed to the people as a dividend payment. In Chile, the withdrawal rules are governed by the Fiscal Responsibility Law of 2006 as part of the fiscal structural balance rule, which limits the maximum amount of withdrawal of resources from the ESSF up to the estimated structural balance deficit that is set as a medium-term objective by the government.

To complement the structural balance rule, the ESSF aims to provide spending stabilization, and reduce dependence on global business cycle and revenue fluctuation due to copper price volatility by allowing government to draw from the Fund to finance fiscal deficits and reduce public debt. This was exactly what the ESSF did until 2014. In 2000–2004, copper revenues accounted for about 5% of total revenue (about 1% of GDP). In the following five years, copper revenues rose sharply to about 24% of total government revenues (almost 6% of GDP) (Sánchez 2011). The ESSF consequently received contributions in 2007, 2008, 2010, 2012, and 2013. The structural balance was estimated as 1% surplus of GDP during the period of 2001–2007, and there was no need to withdraw resources from the ESSF. For 2008, the structural balance was reduced to 0.5% fiscal surplus of GDP and then increased a structural deficit to 1% of GDP as the global financial crisis started unraveling. Some partial withdrawals were made to finance the 2009 fiscal stimulus plan and to finance part of the contribution to the PRF in 2010 and 2014 (Table 1).

In other words, the successful implementation of the robust structural balance rules enabled substantial savings during the copper price boom and contributed to the reduction of public debt, favorable financial conditions in international capital markets for Chile, macroeconomic stability, and a more effective counter-cyclical use of both fiscal and monetary policy. The ESSF was managed more conservatively than other saving funds, such as most oil funds in the Gulf states, and savings and Stabilization Funds, such as the Government Pension Fund Global in Norway, because it invests more in safe and liquid assets that can be used at short

Table 1 ESSF's Contributions and withdrawals (US$ million)

Year	*Contribution*	*Withdrawal*
2007	13,100	
2008	5,000	
2009		9,278
2010	1,362	150
2011		
2012	1,700	
2013	603	
2014		499
2015		464
Total	21,766	10,391

Source https://old.hacienda.cl/english/sovereign-wealth-funds/annual-report.html

notice for intervention purposes when unanticipated shocks hit resource-dependent economies (Alsweilem et al. 2015). The assets of the ESSF grew in a healthy manner, from its initial assets of $2.58 billion in 2007 to $14.7 billion in early 2016. More importantly, the clearly articulated rules in place prevented the government from completely emptying the fund, as had happened in Venezuela.

Many studies on sovereign wealth funds (SWFs), including natural resource funds, rank Chile's funds "among the most transparent in the world with their clear and simple governance structures" (Alsweilem et al. 2015). Both the OECD and the IMF have emphasized the tradition of rule-based macroeconomic and fiscal management in Chile. Pursuing common rules on SWFs was one of the government's objectives when SWFs attracted so much concern in 2007–2008 that the USA and a few European countries threatened to close their doors to SWF investments. Under such pressure and threats, in October 2007 and then in April 2008, countries that owned and hosted SWFs agreed at the International Monetary and Financial Committee to negotiate, among SWFs whose home countries were members of the IMF, a set of rules regulating their operations and activities in return for an open door to investment. The Chilean government offered to host the negotiations, serviced by the International Monetary Fund, that led to the adoption of the voluntary agreement on the 24 Generally Accepted Principles and Practices (GAPP), known as the Santiago Principles (Truman 2010). The 23 SWFs from 21 countries eventually became members of the International Forum of the Sovereign Wealth Funds (IFSWF), which took on the responsibility of overseeing the implementation of the GAPP.

The Santiago Principles cover three main areas:

- Legal framework, objectives, and coordination with macroeconomic policies;
- Institutional framework and governance structure;
- Investment and risk management framework.

The principles have four guiding objectives:

- To help maintain a stable global financial system and free flow of capital investment;

- To comply with all applicable regulatory and disclosure requirements in the countries in which they invest;
- To invest on the basis of economic and financial risk and return-related considerations;
- To have in place a transparent and sound governance structure that provides for adequate operational control, risk management, and accountability.

The ESSF was ranked high in accountability and transparency, higher than the Alaska Permanent Fund and the Future Fund in Australia on the SWF Scoreboard, but not in overall categories of GAPP (Truman 2017). This is, to a large extent, because the ESSF is part and parcel of Chile's budget and fiscal policymaking, and it is therefore accountable to the government. The hierarchical and elite decision-making process ensures limited input from the public at a time when people have argued that the Parliament needs to develop its capacity to scrutinize government economic policies. This is not atypical; many SWFs are managed by experts and professionals with limited oversight, primarily because of the lack of knowledge and expertise of politicians. In Chile, however, the three instances of government attempts to withdraw resources from the ESSF demonstrates that some parliamentarians have gradually become more assertive in making their voices heard, and that there could be meaningful debates when national wealth is concerned.

What is interesting about the ESSF in terms of regulating the behavior of SWFs is that the Chilean government was active in developing the Santiago rules, including hosting and supporting the negotiations on the GAPP. After the working group of SWFs was institutionalized as the International Forum of the SWFs (IFSWF) in 2010, the ESSF and PRF were also active participants in implementing the GAPP and completing their self-assessments. Yet, when the IFSWF decided to relocate from Washington, DC to London and registered as a nonprofit organization in 2014, Chile decided to pull out, as did the Government Pension Fund Global (GPFG) of Norway. The Ministry of Finance in Norway announced that it withdrew the GPFG from the IFSWF because the rest of the member SWFs had not made sufficient progress in transparency and accountability. The Chilean government never explained its position in public; the PRF and the ESSF simply stopped participating in the IFSWF. One speculation is that, since the two funds are part and parcel of the national economic management framework, the Ministry of

Finance both owns and manages them. It is therefore difficult for these investment funds to claim to be "independent" as many other SWFs do. In practice, however, those who advise on the investment and uses of the funds have developed different perceptions and expectations of their roles and responsibilities. This is clearly demonstrated in the following cases.

The Use of the ESSF: Three Case Studies

The Fiscal Responsibility Law of 2006 provides clearly stated rules on sources of funding, deposit rules, and withdrawal rules. Yet, the decision on whether, when, and how much is to be withdrawn is always political, however clear the rules may be. The following three cases outline the actors involved, the arguments of the government, its critics, and the outcome.

2009: The "Successful" Use of the ESSF to Navigate the GFC

Michelle Bachelet was first elected as Chile's first female President in 2006. Despite her high level of popularity, her first government (2006–2010) faced many political and economic challenges, among which were widespread protests against the continuation of the neoliberal economic policies put in place by the previous "Concertación" governments. The massive student demonstrations and workers' strikes in 2006 became known as the "Penguin revolution" (Chovanec and Benitez 2008). Just one month after taking office, the government faced over a million high school students on the streets protesting the privatization of the educational system; workers demanding better job security and work protection; and the general public protesting the "Transantiago" transport system that hurt the most vulnerable part of society (Ruiz 2012).

In 2008, halfway into her first term, Michelle Bachelet reshuffled her cabinet and announced a series of major policy changes, including another round of pension reforms to introduce a minimum pension guarantee and government guarantees of borrowing by state-owned enterprises (Funk 2009; Sánchez 2011). The global financial crisis and the decrease in world copper prices led to a decline in government revenue and GDP growth in 2008 and 2009. The overall fiscal balance reversed from the positive territory to a deficit of 4.4% of GDP in 2009 (Table 2).

The structural balance rule worked according to its design: the fiscal surplus rose from 2% in 2004 to over 7% of GDP in 2007, allowing Chile

Table 2 Chile's Fiscal Balances

	2001	*2002*	*2003*	*2004*	*2005*	*2006*	*2007*	*2008*	*2009*
Overall balance	−0.5	−1.2	−0.2	2.1	4.6	7.7	8.8	5.2	−4.4
Structural balance	1.0	0.7	1.0	1.0	1.0	1.0	1.0	0.6	−1.7
Real expenditure growth	5.8	4.7	2.1	5.3	6.7	4.9	8.4	7.8	17.8

Source Sánchez (2011)

to save more than 10% of GDP in its investment funds. With the impact of the global financial crisis, this surplus disappeared. The government decided to withdraw from the ESSF to finance the deficit and support its expansionist policies, as many economies did at the time. In January 2009, Minister of Finance Andrés Velasco announced a fiscal stimulus plan of $4 billion, equivalent to about 3% of GDP (Parrado 2017), aiming to increase fiscal expenditure and stabilize the economy (Economía y Negocios 2009). In June, Velasco announced a withdrawal of an additional $4 billion from the ESSF to boost the economy, stating that during the economic downturn, a higher fiscal expenditure was "indispensable" to economic stability:

> Using accumulated savings and leveraging our ability to issue debt in local markets, where we have practically not issued debts, allow us to spend more this year. These measures are absolutely necessary to inject more dynamism and greater energy into the economy. (Martínez 2009)

Most political and economic actors, including the opposition in Parliament, supported the government decision to use the ESSF and apply counter-cyclical measures. The almost unique exception was the National Agricultural Society, which was concerned about the potential negative effects of state intervention in the agriculture sector's competitivity (El Mercurio de Valparaíso 2009). To avoid criticism, Minister Velasco emphasized that the decision was made because of economic necessity, not because of pressure from specific electoral groups (Economía y Negocios 2009).

By the end of 2009, $9.278 billion had been withdrawn from the ESSF to finance an expansionary fiscal policy that included public investment, employment and credit subsidies, and the capitalization of state companies and banks, such as the $2.278 billion capitalization of CODELCO, the main state copper company, the doubled capitalization of Banco Estado, the only public bank of Chile, and the main hydrocarbon company, ENAP. The expansionary fiscal policy also covered tax reductions in strategic sectors and credit transfers to others (Chilean Ministry of Finance 2009; Ffrench-Davis 2016; Parrado 2017). The policy proved successful in mitigating the negative effects of the global financial crisis.

Despite the global financial crisis and a huge reconstruction bill after the 2010 earthquake and tsunami, Chile was able to maintain an average annual GDP growth rate of 4.6% in comparison with the OECD average of 1.9% in 2009–2014. The strengthened fiscal rules, higher tax revenues to finance long-term spending increases, and structural balance in general served the Chilean economy well with low national financial debt of 15.1% of GDP in comparison with 112.6% of OECD average in 2014 (OECD 2015). The government started replenishing the PRF and the ESSF in the last quarter of 2011. At the same time, the robust fiscal situation, notably the near-absence of net debt, provided the government some space to push forward its new "Ingreso Ético Familiar" Programme, which targeted direct cash transfers to families living in extreme poverty both to bolster their income and to improve their productivity and contribution to the broad economy (OECD 2012).

Chile's economic performance was highly praised by the OECD and the IMF for its efficient use of the ESSF to navigate a global economic slowdown, and more broadly for its institutional framework buttressed by the Fiscal Responsibility Law of 2006. The ESSF and the structural balance rule allowed the government to adopt an assertive fiscal policy and expand fiscal expenditure, with its GDP growth back to nearly 6% in 2010–2012 while most OECD countries were struggling to get out of the recession. Politically, the domination of the Minister of Finance, along with its Financial Committee, and the Central Bank was further legitimized and strengthened. The narrative promoted by the government that economic policies should be left to technical professionals prevailed. Since 2009, however, the use of the ESSF has been highly contested among several political and economic actors, outside and within the government, as the next two cases show.

2014: The Debate Around the Use of the ESSF to Capitalize CODELCO

The second occasion of the Chilean government's proposal to withdraw resources from the ESSF was in 2014, after Michelle Bachelet was elected as President for the second time. The last two years of the previous government, led by the center-right coalition under President Sebastián Piñera, had experienced some economic difficulties. The world copper price started falling in early 2012, alongside a decline in demand in China. By then, Chile produced over one-third of the world's copper supply, and copper represented about half of Chile's total exports. The destinations of its exports had changed significantly in the previous decades: exports to Asia accounted for over 46% of its total export with 23% going to China alone, while exports to North America (Canada, the USA, and Mexico) amounted to only 12%. Nonetheless, Chile's economy remained resilient, with a GDP growth rate of 6% at the beginning of 2013, in contrast to 3% in other Latin American countries. Unemployment remained stable, at a relatively low rate of 8.5% in early 2012, down to 5.6% in mid-2013, and back to 6.3% a year later (OECD 2015). The resilient economic development was due, to a large extent, to the sound macroeconomic policies, a robust fiscal framework, and a flexible exchange rate regime.

Soon after taking office in 2014, Michelle Bachelet announced her government's five-year investment strategy, "Plan de Negocios y Desarrollo 2014–2018," that included seven major projects with a cost of $12.265 billion (Chilean Ministry of Finance 2014b). One of the components of the plan was to inject more capital to CODELCO. In April 2014, in an interview with Bloomberg, the newly appointed undersecretary of finance, Alejandro Micco, announced that the government was considering withdrawing resources from the ESSF to finance the capitalization of CODELCO (IIMCh 2014). The minister of finance later confirmed it, arguing that "the government believes in a strong CODELCO" because it is "the heritage of all Chileans; it is a very important company in many dimensions," and that "If CODELCO did well, Chile would do well." The government was committed to supporting CODELCO for three main reasons: world copper prices started falling after 2012 and seemed likely to remain low for the coming years; costs of production were rising; and the cost of borrowing in international financial markets was surging. The chief executive of CODELCO, Thomas Keller, immediately publicly endorsed the government's proposal to use the resources of the ESSF to capitalize the company:

> We can celebrate this initiative that is aimed at providing the corporation with the necessary resources so that we can continue to implement the most important investment program in the history of the company, which allows CODELCO to be projected as the main company of Chile, the main producer of surpluses for the State. (América economía 2014)

The issue of withdrawing funds from the ESSF to refinance the state-owned copper corporation became a highly contested issue within the government and among the public for several reasons. First, in 2014, although the economy had slowed down, Chile was not facing as serious an economic or financial crisis as it did in 1998–1999 during the Asian financial crisis or in 2008–2009 during the global financial crisis. Indeed, the economy in Chile performed better than other economies in Latin America, with a higher GDP growth rate and lower unemployment rates. The central government debt grew from about 8% of GDP in 2010 to 15% in 2014 but remained low compared to those of its neighbors and of other OECD countries. The total wealth of the PRF and the ESSF was growing with replenishment. For the opposition, the economic conditions did not justify the withdrawal of resources from the ESSF. Second, the opposition had become more vocal in contesting government policies. By 2014, democracy in Chile had matured, having successfully gone through government transitions from the center-left to center-right and back to center-left. The opposition had had its opportunity to govern and would be able to do so again if it could rally sufficient support for the next election. Government agencies and key economic institutions had also become more representative. The two key elements in fiscal institutions were the "output gap" panel and the "copper price" panel, each of which consisted of 15 well-known economists from academia, research institutes, and industries, appointed by the Minister of Finance, from both governing and opposition coalitions (Vammalle and Rivadeneira 2017). The Financial Committee was similarly composed of members from both major political coalitions. These committees were expected to provide non-political and independent advice based on which the minister would make decisions.

In 2014, this practice was challenged. Arturo Cifuentes, the chair of the Financial Committee, an independent body of experts that advised the minister of finance on the long-term investment policy of the PRF and the ESSF and other related issues, was one of its most outspoken critics: "I was surprised. First, because the Financial Committee has never studied this possibility, and second, because the idea goes against all our recommendations" (Pulso 2014).

For Cifuentes, the ESSF was not only a Stabilization Fund, but also a savings fund. At a time when copper prices were low, the ESSF needed to diversify its investment to assets "whose behaviour is opposite to copper … an important part of the work of the Financial Committee, in recent years, has been oriented to see how to diversify." In addition, he argued that the government had not consulted with the Financial Committee that would have advised differently: "I have been on the Committee for almost three years and the possibility of using these funds to help CODELCO or another state firm, has never been discussed" (Pulso 2014).

Cifuentes' public attack on the government plan was unprecedented. He spoke out not only as an individual, but as a member of the Financial Committee that is considered a technical professional body. The Committee is meant to provide apolitical advice to the minister of finance. Then-Finance Minister Alberto Arenas did not appreciate the criticism and decided to fire Cifuentes from his position as chair of the Financial Committee. In response, Cifuentes refused to resign, arguing that his position was technical and not political, and the minister's decision to fire him was intended to undermine the Committee: "If people are going to be taken out of the committee every time they have divergent opinions, what is the use of it?" (Weissman 2014). The minister directed the Office of the Comptroller General of the Republic to fire Cifuentes anyway and replaced him with José de Gregorio, another well-known economist and former governor of the central bank of Chile from 2011–2017 (El Mostrador 2014). Adding to this saga, the governor of the Central Bank also challenged the government's proposal to withdraw resources from the ESSF to finance the capitalization of the state copper company, arguing that the economic conditions in 2014 did not warrant the withdrawal:

> They were used in 2009, correctly used, because it was precisely a period of the lowest global growth since the Great Depression, and we have accumulated a significant amount once again, which we must keep and use in these circumstances, not in others. (El Mercurio 2014)

Facing strong opposition from Chile's two key economic institutions, and in the absence of widespread public support, the government abandoned the plan. Instead, in October 2014, the government proposed in Parliament to raise $4 billion to help capitalize CODELCO, arguing that "it would be tremendously irresponsible to stand idly, watching the global status of CODELCO threatened and its contribution to national revenue and the welfare of the people diminish" (24 Horas 2014). The Parliament passed the bill and CODELCO received government money to capitalize itself.

The case shows that, as in the case of most sovereign wealth funds, decisions concerning the PRF and ESSF are often made among a small group of professional elites, without drawing much public attention. Yet, such decisions can be controversial because of the complexities in managing and using the ESSF.

2016: Another Announcement of the Use of the ESSF, Another Back Down

The third case involving the use of funds from the ESSF took place in July 2016, when Finance Minister Rodrigo Valdés indicated that the government was considering the possibility of using the ESSF to balance the 2017 budget. Having learned from the 2014 incident, the government cautiously adopted a different strategy to avoid the high internal and external contestation that had blocked the government's action and tarnished the reputation of the ministry as a professional institution. This time, Valdés chose a public and official forum to make the announcement. In early July 2016, at a seminar organized by two powerful players in Chilean economy, the Federation of Chilean Industry (Sofofa, its name in Spanish) and the Santiago Chamber of Commerce (Chilean Ministry of Finance 2016), Valdés explained the challenging fiscal situation and the reason behind exploring the use of SWFs to finance the deficit in the following terms:

> Fiscal expenditure in Chile is determined by a fiscal rule [...] we have a relatively significant deficit this year and next year we will also have a deficit. As in any company or any family that spends more than the income it has, it must finance that part. That part is usually financed with debt, but we need to ask ourselves if it is only possible by issuing debts or we could use some part of the assets that we have. (Chilean Ministry of Finance 2016)

The issue was, nonetheless, quickly politicized. Various parties within the coalition supported the government's initiative to spend in order to create jobs, but they did not agree whether withdrawing the resources of the ESSF was the best way to finance the budget deficits (Nuevo Poder 2017). The issue became politically controversial when the Chilean Communist Party sent a letter to Valdés, indicating that it not only supported but also demanded the government use the "assets that we have" to finance the fiscal expenditures. The opposition responded immediately: "If the PC [Communist Party] thinks that more can be spent using sovereign wealth funds, it has not understood basic accounting principles," Andrés Velasco, former Minister of Finance, argued on social media, criticizing the very discussion of the possibility of withdrawing from the ESSF for deficit spending (El Dínamo 2016).

The debate over the budget was normal and healthy; indeed, it covered a range of issues in addition to the question of withdrawal from the ESSF. Among these issues were the deceleration of economy, the copper prices that were lower than what had been calculated in the previous year, and the rising demand for pension and social security payments. In October, before the tensions could escalate to the levels of 2014, the government decided to abandon the plan to withdraw more resources from the ESSF in order to avoid derailing other reforms on the agenda. The minister said that, given the external and internal economic conditions, the government would be able to increase its debt level to finance the budget deficits instead of withdrawing from the ESSF (EMOL 2017). Two things stood out in this debate: the government tried to keep the discussion at the technical level, and it took a cautious approach by consulting its expert advisory committees.

Even though the rules of (International Forum of Sovereign Wealth Funds 2006) stipulate that withdrawals from ESSF can be made at the discretion of the Finance Minister in order to fund the PRF, cover shortfalls in government revenue overspending, or reduce the national debt, recent proposals to use the assets of the ESSF became quite controversial. In 2014, for example, the Ministry of Finance calculated the target structural balance as a 1% deficit, indicating the need for withdrawals from the ESSF, rather than deposits, which are required when there is a structural surplus. The proposal was killed not so much because of the different views on how to use the resources, but rather because of the lack of consultation with expert panels. Furthermore, as its democratic system matured, especially after the 2015 Constitutional reform that reduced the term of the President from six years to four, different ideas are now more freely put in public. As the political system developed, parliamentarians became more confident in speaking out, as both the center-left and the center-right had opportunities to govern, and the parliamentarians gained more expertise.

Conclusion: Is Chile a Successful Case of Management and Use of SWFs?

Chile has been highly praised for its commitment to fiscal discipline and its rule-based governance, concentrated in "highly respected and trusted institutions" such as the Chilean Directorate of Budget, the independent expert panels in calculating the structural balance—the trend of GDP growth and the evolution of copper price—based on which the government decides its fiscal policy. The PRF and the ESSF are at the center of "fiscal sustainability, economic growth and reduction of extreme poverty" (Vammalle and Rivadeneira 2017). ESSF is one of the good examples, serving as a counter-cyclical fiscal tool to insulate the economy from internal and/or external shocks, such as changes in commodity prices, to smooth consumption.

This chapter provides an overview of the Chilean SWFs following the official "success story" narrative, which describes the Chilean case as one in which the government learned from the mistakes it had made during the Asian crisis in the 1990s, established a strong institutional framework to maintain macroeconomic balance via a strict structural rule and the Fiscal Responsibility Law, and effectively managed and used its SWFs to navigate the most challenging rainy season of the past few decades—the GFC—by applying counter-cyclical measures with their funds. Having a small economy highly dependent on one single commodity (copper), Chile is presented as a "textbook" example of effective use of SWFs to diminish the exposure of a single-source economy to external shocks related to the fluctuations of commodity prices.

The discussion also goes beyond the official story to examine three cases in which the SWFs were either used or potentially used, in order to determine whether Chile was indeed a "textbook" case of proper management and use of SWFs. Many conclusions can be drawn from this exercise. The first and most relevant conclusion is that Chile remains a success story not only in the region but among countries with SWFs as a whole. If we consider the basic "textbook" characteristics that a proper management should follow in terms of securing future pensions (PRF) and stabilizing the economy (ESSF), the Chilean government has used both SWFs to counter the negative effects of the GFC, while avoiding politicization and continuing to allocate funds for future rainy seasons.

The second conclusion is that the reality is much more complex than the official story, which was derived mostly from the 2009 experience. The other two case studies have raised questions. After 2009, the government tried to use the funds in 2014, and hinted at the option of a second attempt in 2016, facing significant contestation from within and outside the government in the former, leading to unprecedented public debates that tarnished the "technical" image of the Ministry of Finance and even led to the dismissal of the President of the Financial Committee, the main institution in charge of advising on the use and management of the SWFs. In 2016, the government had to be more careful, and ultimately desisted in order to avoid criticism.

A third conclusion that can be drawn from this analysis is that two factors play major roles in the successful use of SWFs that are meant to maintain macroeconomic stability: an evident crisis or rainy season, and the cautious and technical management of the government's narrative. In 2009 the contagion of the GFC and its significant potential negative implications were evident, and the government's narrative followed an explicitly technical approach that was backed by every sector of the political and economic system. In 2014, the economy was slowing down, but the situation was far from that in 2009, and the government managed its narrative poorly, opening the door for criticism. In 2016, the government managed its narrative more effectively but lacked good arguments to support the use of the ESSF, as once again there was no clear evidence of a "rainy season" at the levels of 2009.

References

24 Horas (2014) Presidenta Bachelet promulga ley sobre aumento de capital a Codelco. 24 Horas, 24 Oct. https://www.24horas.cl/economia/presidenta-bachelet-promulga-ley-sobre-aumento-de-capital-a-codelco-1468583

Aggarwal VK (2000) Exorcising Asian debt: Lessons from Latin American rollovers, workouts, and writedowns. In: Private capital flows in the age of globalization: The aftermath of the Asian crisis. Edward Elgar, Northampton, p 105–139

Alsweilem KA, Cummine A, Rietveld M, Tweedie K (2015) A comparative study of sovereign investor models: Sovereign fund profiles. Harvard Kennedy School, Belfer Center For Science and International Affairs

América economía (2014) Codelco respalda opción de capitalizar a la estatal con fondos soberanos. América economía, 15 Apr. https://www.americaeconomia.com/negocios-industrias/codelco-respalda-opcion-de-capitalizar-la-estatal-con-fondos-soberanos

Angell A (1993) The transition to democracy in Chile: A model or an exceptional case? Parliamentary Affairs 46(4):563–579

Bacon R, Tordo S (2006). Experiences with oil funds: Institutional and financial aspects. Energy Sector Management Assistance Program, Washington, DC. https://openknowledge.worldbank.org/bitstream/handle/10986/17952/ESM3210Experiences1Oil1Fund01PUBLIC1.pdf?sequence=1&isAllowed=y

Bauer A, ed., (2014) Managing the public trust: How to make natural resource funds work for citizens. Revenue Watch Institute, New York. https://www.resourcegovernance.org/sites/default/files/NRF_RWI_Complete_Report_EN.pdf

Blöndal JR, Curran J (2004) Budgeting in Chile. OECD. https://www.oecd.org/chile/40139802.pdf

Brender V (2010) Economic transformations in Chile: The formation of the Chicago Boys. The American Economist 55(1):111–122

Chilean Ministry of Finance (2008) Annual Report Sovereign Wealth Funds 2008. DIPRES. https://www.hacienda.cl/english/sovereign-wealth-funds/annual-report/annual-report-sovereign-wealth-funds-748.html

Chilean Ministry of Finance (2009) Annual Report Sovereign Wealth Funds 2009. DIPRES. https://www.hacienda.cl/english/sovereign-wealth-funds/annual-report/annual-report-sovereign-wealth-funds.html

Chilean Ministry of Finance (2010) Annual Report Sovereign Wealth Funds 2010. DIPRES. https://www.hacienda.cl/english/sovereign-wealth-funds/annual-report/annual-report-sovereign-wealth-funds-762867.html

Chilean Ministry of Finance (2014a) Annual Report Sovereign Wealth Funds 2014. https://www.hacienda.cl/english/sovereign-wealth-funds/annual-report/annual-report-sovereign-wealth-funds-506958.html

Chilean Ministry of Finance (2014b) Capitalización de Codelco en Cifras. https://www.hacienda.cl/sala-de-prensa/archivo-2014-2018/infografias/capitalizacion-de-codelco-en-cifras.html

Chilean Ministry of Finance (2015) https://old.hacienda.cl/english/sovereign-wealth-funds/annual-report.html

Chilean Ministry of Finance (2016) Ministro Valdés afirma que "está garantizado que Chile va a llegar a la gratuidad universal" si se aprueba el proyecto de ley de educación superior. https://www.hacienda.cl/sala-de-prensa/archivo-2014-2018/noticias/ministro-valdes-afirma-que-esta.html

Chovanec DM, Benitez A (2008) The penguin revolution in Chile: Exploring intergenerational learning in social movements. Journal of Contemporary Issues in Education 3(1):39–57

Cumming DJ, Wood G, Filatotchev I, Reinecke J (2017) The Oxford handbook of sovereign wealth funds. Oxford University Press, Oxford

De Mesa AA, Mesa-Lago C (2006) The structural pension reform in Chile: Effects, comparisons with other Latin American reforms, and lessons. Oxford Review of Economic Policy 22(1):149–167

DeShazo P (2005) Chile's road to development, 1990–2005: Lessons learned. https://csis-website-prod.s3.amazonaws.com/s3fs-public/legacy_files/files/media/csis/pubs/pp_0508deshazo.pdf

Economía y Negocios (2009) Hacienda justifica la cuantía del plan récord de estímulo fiscal y rebaja su proyección de IPC. Economía y Negocios, 7 Jan. http://www.economiaynegocios.cl/noticias/noticias.asp?id=58199

El Dínamo (2016) Velasco dice que el PC "no entiende contabilidad" por pedir que el Estado aumente gasto fiscal con fondos soberanos. El Dínamo, 24 Aug. https://www.eldinamo.cl/nacional/2016/08/24/velasco-dice-que-el-pc-no-entiende-contabilidad-por-pedir-que-el-estado-aumente-gasto-fiscal-con-fondos-soberanos/

El Mercurio (2014) Codelco promueve capitalización con fondos soberanos realzando retornos de la minería. El Mercurio, 18 July. http://zublin.icc-crisis.com/?p=5376

El Mercurio de Valparaíso (2009) Aumentarán gasto para reactivar economía. El Mercurio de Valparaíso, 17 June. http://www.mercuriovalpo.cl/prontus4_noticias/site/artic/20090617/pags/20090617000419.html

El Mostrador (2014) José De Gregorio finalmente llega a Hacienda. El Mostrador, 2 Sept. https://www.elmostrador.cl/mercados/sin-editar-mercado/2014/09/02/jose-de-gregorio-finalmente-llega-a-hacienda/

EMOL (2017) Hacienda estima PIB de 2,25% para 2017 y descarta uso de fondos soberanos para financiar déficit. EMOL, 3 Oct. https://www.emol.com/noticias/Economia/2016/10/03/824759/Hacienda-se-alinea-con-el-Banco-Central-y-preve-que-la-economia-crecera-225-en-2017.html

Engel EM, Galetovic A, Raddatz CE (1999) Taxes and income distribution in Chile: Some unpleasant redistributive arithmetic. Journal of Development Economics 59(1):155–192

Fernandez A, Vera M (2012) The Bachelet Presidency and the end of Chile's Concertacion era. Latin American Perspectives 39(4):5–18

Ffrench-Davis R (2003) Chile, entre el neoliberalismo y el crecimiento con equidad. Nueva Sociedad 183:70–90

Ffrench-Davis R (2016) La experiencia de Chile con el balance fiscal estructural. Cuadernos de economía (Santafé de Bogotá) 35(67):149–171

Fourcade-Gourinchas M, Babb SL (2002) The rebirth of the liberal creed: Paths to neoliberalism in four countries. American Journal of Sociology 108(3):533–579

Funk RL (2009) Chile: segundo tiempo. Revista de ciencia política (Santiago) 29(2):301–326

Iglesias-Palau A (2009) Pension reform in Chile revisited: What has been learned? https://doi.org/10.1787/224473276417

IIMCh (2014) Gobierno se abre a opción de usar fondo soberano para capitalizar a Codelco. Diario Financiero, 15 Apr. https://www.iimch.cl/index.php/noticias2/actualidad/738-gobierno-se-abre-a-opcion-de-usar-fondo-soberano-para-capitalizar-a-codelco

International Forum of Sovereign Wealth Funds (2014) Santiago Principles: 15 case studies. November. https://www.ifswf.org/sites/default/files/SantiagoP15CaseStudies1_0.pdf

Lienert I (2010) Should advanced countries adopt a fiscal responsibility law? International Monetary Fund.

Martínez R (2009) Gasto fiscal, imprescindible para apoyar economía Chile:ministro. Reuters, 17 June. https://lta.reuters.com/articulo/negocios-economia-chile-gobierno-idLTASIE55F12U20090616

Nuevo Poder (2017) NM pide evaluar uso de fondos soberanos para aumentar inversión. Nuevo Poder, 3 Oct. http://www.nuevopoder.cl/nm-pide-evaluar-uso-de-fondos-soberanos-para-aumentar-inversion/

Ocampo JA (2014) The Latin American debt crisis in historical perspective. In: Life after debt. Springer, p 87–115

OECD (2003) OECD economic surveys: Chile. https://www.oecd-ilibrary.org/economics/oecd-economic-surveys-chile-2003_eco_surveys-chl-2003-en

OECD (2010) OECD economic surveys: Chile. https://www.oecd-ilibrary.org/economics/oecd-economic-surveys-chile-2010_eco_surveys-chl-2010-en

OECD (2012) OECD economic surveys: Chile. https://www.oecd-ilibrary.org/economics/oecd-economic-surveys-chile-2012_eco_surveys-chl-2012-en

OECD (2015) OECD economic surveys: Chile. https://www.oecd-ilibrary.org/economics/oecd-economic-surveys-chile-2015_eco_surveys-chl-2015-en

Parrado E (2017) Is it possible to avoid the St Augustine syndrome of fiscal procrastination? The case of Chile. In: Cumming D, Wood G, Filatotchev I, Reinecke J (eds) The Oxford handbook of sovereign wealth funds. Oxford University Press, Oxford

Pulso (2014) Comité Asesor de los Fondos Soberanos sobre Codelco: «Es la peor decisión que se puede tomar». Pulso, 17 Apr. http://www.mch.cl/2014/04/17/comite-asesor-de-los-fondos-soberanos-sobre-codelco-es-la-peor-decision-que-se-puede-tomar/

Rodriguez JC, Tokman CR, Vega AC (2007) Structural balance policy in Chile. OECD Journal on Budgeting 7(2):59–92

Ruiz C (2012) New social conflicts under Bachelet. Latin American Perspectives 39(4):71–84

Saffirio C (2013) Regulación de fondos soberanos: El caso chileno. Anuário de Derecho Público, UDP:569–584

Sánchez TD (2011) Strengthening Chile's rule-based fiscal framework. International Monetary Fund

Schmidt-Hebbel K (2012) Fiscal institutions in resource-rich economies: Lessons from Chile and Norway. Citeseer

Siavelis PM (2009) Enclaves de la transición y democracia chilena. Revista de ciencia política (Santiago) 29(1):3–21

Truman E (2010) Sovereign wealth funds: Threat or salvation? Peterson Institute.

Truman EM (2017) Sovereign wealth fund transparency and accountability explored. Wake Forest Law Review 52:997

Undurraga T (2015) Neoliberalism in Argentina and Chile: Common antecedents, divergent paths. Revista de Sociologia e Política 23(55):11–34

Valero D, Monterde A (2011) Managing investment risk in Chilean pension funds. OECD.org

Vammalle C, Rivadeneira AMR (2017) Budgeting in Chile. OECD Journal on Budgeting 16(3):9–70

Weissman I (2014) Culebrón inédito en Hacienda: Arenas saca al Presidente del Comité Financiero por criticar uso de fondos soberanos en Codelco y pone a De Gregorio. El Mostrador, 4 June. https://m.elmostrador.cl/mercados/2014/06/04/culebron-inedito-en-hacienda-arenas-saca-al-presidente-del-comite-financiero-por-criticar-uso-de-fondos-soberanos-en-codelco-y-pone-a-de-gregorio/

Saving Today's Bread for Tomorrow's Consumption? The Politics of Trade-offs in the Governance of Ghana's Petroleum Funds

Ishmael Ackah and Denis M. Gyeyir

Introduction

Experience is often the best teacher, so the adage says. Indeed, learning from one's experience and those of others is a good practice. To this end, attempts were made to learn lessons from the mining sector and the experiences of other countries when Ghana discovered oil in commercial quantities in 2007. Indeed, for over more than a century, revenue from Ghana's mineral resources has not been guided by a framework that spells out criteria for allocation, disbursement, and use of mineral revenues (Ackah et al. 2020; Standing 2014). The enactment of the Minerals Development Fund Act, 2016 (Act 912), which provides guidance for the application of 20% of mineral royalties for the benefit of mining communities, traditional authorities, local government, and land interest holders, was a recent attempt at doing so. The larger mineral revenues remain

I. Ackah (✉)
Ghana Institute of Management and Public Administration, Accra, Ghana

D. M. Gyeyir
Natural Resource Governance Institute (NRGI), Accra Office, Ghana

E. Okpanachi and R. Chowdhari Tremblay (eds.), *The Political Economy of Natural Resource Funds*, International Political Economy Series,
https://doi.org/10.1007/978-3-030-78251-1_6

subject to utilization under the central government's budgetary arrangements, with attendant challenges (Tsikata 1977; Alagidede and Akpoza 2012). In the cocoa sector, revenues are equally treated as consumption capital without explicit fiscal provisions for savings or investment. According to Gyeyir (2019), the Cocoa Stabilization Fund, established in 2015, serves as a risk mitigation mechanism for farmers with a primary object of sustaining their earnings in seasons of world cocoa price downturns.

The discovery of oil, however, presented different challenges, which can be categorized into 4 main strands. First, there was much expectation from both politicians and the public (Andrews 2013). Indeed, the then-President popped champagne on live television and declared, "We're going to really zoom, accelerate ... and you'll see that Ghana truly is the African tiger" (*BBC News* 2007). To some communities, policymakers, and individuals, oil was going to solve all of Ghana's problems. Second, there were institutions, mainly in civil society and academia, who, having read the experiences of oil resource management, especially in Africa, were cautiously pessimistic (Bawumia and Halland 2017; Osei-Tutu 2012; Gyampo 2010). According to this group, looking at the experiences of the mining sector and how oil has contributed to poverty, Dutch Disease, and conflicts in other countries in Sub-Saharan Africa, only a miracle could make Ghana's case different. Third, the popular opinion was that with the right set of laws and regulations, and following the examples of countries such as Norway, Malaysia, and Trinidad and Tobago, oil would be the anchor for national development (Graham et al. 2016). However, Ackah et al. (2020) have concluded that unless the laws are effectively implemented, tracked, and reviewed for their effects on the governed, they will just be laws on paper. Indeed, since most of these laws, such as the Petroleum Revenue Management Act, have discretionary provisions, implementers will find ways of going around them. Finally, there were those who raised issues about the potential challenges on how the benefits of the oil production would be shared (Agbefu 2011; Okpanachi and Andrews 2012; Panford 2014; Abraham 2019). Indeed, not long after the discovery of oil, the traditional rulers of the Western Region, where the oil has been discovered, requested 10% of the revenues for the region (Kpodo 2010). This was based on the fact that the Western region provides most of Ghana's natural resources, but faces many developmental challenges. Others advocated for local and community content to ensure that oil translates into direct benefits (Ablo and Asamoah 2018).

Upon the commercial discovery of oil in 2007, Ghana undertook a series of consultations on how to manage proceeds from this new sector. Ghana, learning from its own experience in managing its other resource sectors, and from experiences of other resource sectors and other resource-rich economies, included the Ghana Petroleum Funds in its fiscal regime on petroleum revenue management. Consultations were held in all ten regions of the country at the time. Key issues from these engagements are summarized in the schema in Fig. 1.

The key questions about the establishment of the Natural Resource Fund (NRF) revolve around how much to save, where to save it, and whether savings should be made for intergenerational equity, over which debate still lingers. These debates were informed by the lack of trust in the political establishment to secure savings in country and the potential of these funds to sustain a stable growth of the economy. In addition, civil society and interest groups were concerned about transparency and accountability for allocations, savings, and uses of petroleum revenues.

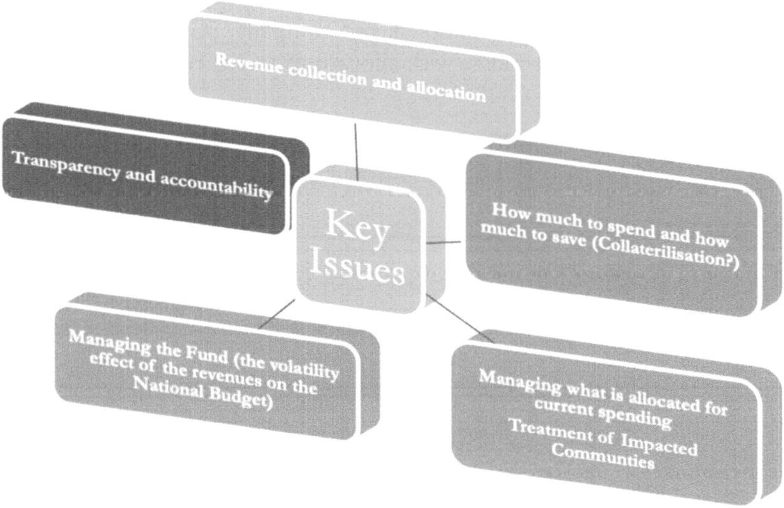

Fig.1 Key issues from consultations

Establishment Stage

Rules and Structures

Based on the consultation, the Petroleum Revenue Management Act (PRMA), 2011 (Act 815), was subsequently enacted "to provide the framework for the collection, allocation and management of petroleum revenue in a responsible, transparent, accountable and sustainable manner for the benefit of the citizens of Ghana." The Act established a holding fund into which all proceeds from petroleum due to the government are deposited. Deposits into the holding fund are net of the national oil company's entitlements for its share of equity participation, as it is mandated to participate in development and production of petroleum revenues on behalf of the government.

Transfers from the Petroleum Holding Fund (PHF) are guided by a formula, specifying in percentage terms, ceilings, and floors, for transfers into the respective destination funds. Figure 2 is a visual representation of the transfer and allocation regime for petroleum revenues under the PRMA.

Under the PRMA, the national oil company (GNPC) receives a percentage of the oil revenues, with allocations to the budget, the Ghana

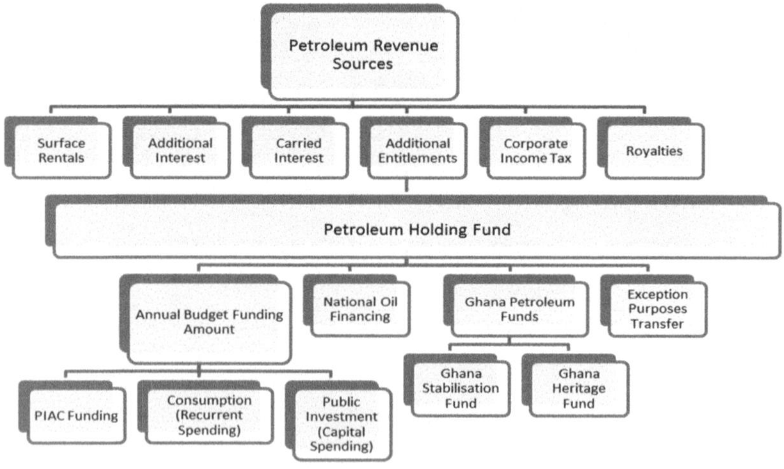

Fig. 2 Fiscal rules under the PRMA (*Source* Aryeetey and Ackah [2018])

Stabilization Fund (GSF), and the Ghana Heritage Fund (GHF). Not more than 70% of net revenues in the PHF, after deduction of the share of the National Oil Company (NOC) is transferred into a Dollar Account of the Annual Budget Funding Amount (ABFA), from whence it is subsequently transferred into a Ghana Cedi Account for spending. The share of the National Oil Company should not be more than 50% of the carried and participating interest and is approved by Parliament. The ABFA is spent in accordance with, but not limited to, four priority areas selected by the Minister of Finance and approved by Parliament every three years in the absence of a long-term national development plan. Twelve priority areas have been specified in the PRMA for selection. These areas are:

a. Agriculture and industry;
b. Physical infrastructure and service delivery in education, science, and technology;
c. Potable water delivery and sanitation;
d. Infrastructure developmentin telecommunication, road, rail, and port;
e. Physical infrastructure and service delivery in health;
f. Housing delivery;
g. Environmental protection, sustainable utilization, and protection of natural resources;
h. Rural development;
i. Developing alternative energy sources;
j. The strengthening of institutions of government concerned with governance and the maintenance of law and order;
k. Public safety and security; and
l. Provision of social welfare and the protection of the physically handicapped and disadvantaged citizens.

Provision is made for spending at least 70% of the ABFA for public investment expenditure. With the amendment of Act 815 in 2015, an infrastructure investment vehicle, the Ghana Infrastructure Investment Fund (GIIF), was indicated to receive a maximum of 25% of the ABFA allocated for public investment expenditure. Provision was also made from the ABFA for funding the activities of the Public Interest and Accountability Committee (PIAC), an additional public oversight body under the Act.

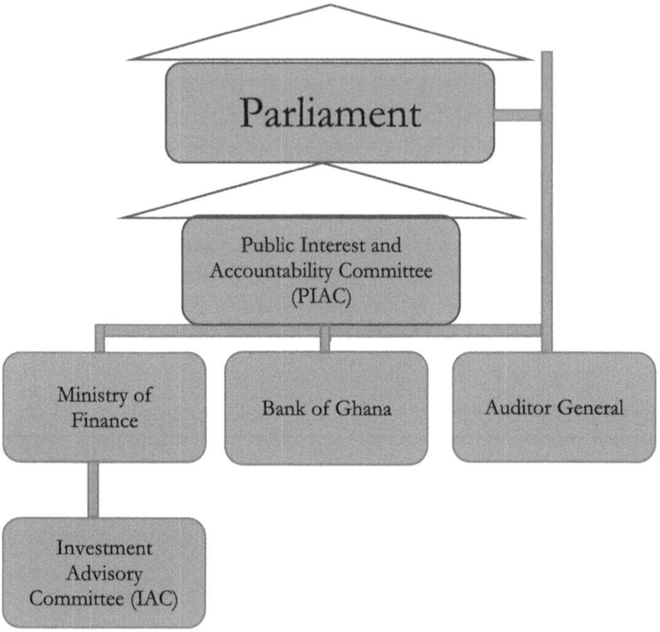

Fig. 3 Structure of institutional oversight on the GPFs

Structure—Institutional Framework and Governance Structure
The PRMA, as amended, assigns clear rules to the management of the GPFs. The institutional setup of the operational management, reporting and disclosures, advisory services, approval, oversight, monitoring, and engendering of public dialogue around the management of the funds have been delineated and are shown in Fig. 3.

Parliament
Parliament lies at the apex of the oversight structure as an arm of government that serves as a check on the exercise of executive power in general and on the management of the GPFs in particular:

- exercises oversight in checking the discretionary power of the Minister of Finance in capping and approval of proposed withdrawals from the GSF and transfer of a portion of accrued interest on the GHF, at 15-year intervals from operation of the Fund;

- receives reports of reconciliation of actual petroleum receipts;
- receives semi-annual reports from the Bank of Ghana and semi-annual and annual reports from the Public Interest and Accountability Committee (PIAC);
- approves the investment policy prepared by the Investment Advisory Committee (IAC) and submitted by the Minister of Finance;
- receives and considers the audit reports and "special" audit reports by the Auditor on the receipts, disbursement, and use of the GPFs;
- receives and considers annual reports of the Minister of Finance on the Petroleum Funds as part of the budget presentation to Parliament;
- approves information and data, the disclosure of which could be inimical to the performance of the GPFs.

The Public Interest and Accountability Committee (PIAC)

This committee is a novelty in the oversight structure and governance of petroleum resources because it consists of nongovernmental institutions such as the Ghana Bar Association, National House of Chiefs, National House of Queens, and others created by the law to oversee petroleum revenue expenditure. The composition and the financing of the PIAC is independent from the government or the public sector. The thirteen-member committee is made up of representatives nominated by independent policy research think tanks, civil society organizations and community-based organizations, the Trade Union Congress, the National House of Chiefs, the Association of Queen Mothers, the Association of Ghana Industries and Chamber of Commerce, the Ghana Journalists Association, the Ghana Bar Association, the Institute of Chartered Accountants, the Ghana Extractive Industries Transparency Initiative, Christian groups, Muslim groups, and the Ghana Academy of Arts and Sciences. The PIAC reports to the citizens of Ghana through direct and indirect engagements and to Parliament through the Finance Committee, usually with representation from Mines and Energy and the Public Accounts Committees. The committee's semi-annual and annual reports are deliberated upon by the Finance Committee and, when appropriate, directives are issued.

The PIAC was established in 2011 under Section 51 of the Petroleum Revenue Management Act (PRMA) (Act 815). The Committee has three main objectives as outlined in the PRMA:

- to monitor and evaluate compliance with the Act by government and relevant institutions in the management and use of petroleum revenues and investments;
- to provide space and platform for the public to debate on whether spending prospects and management and use of revenues conform to development priorities as provided under Section 21; and
- to provide independent assessment on the management and use of petroleum revenues to assist Parliament and the Executive in the oversight and the performance of related functions.

Bank of Ghana

The Bank of Ghana performs the following functions:

- Undertakes operational management of the GPFs under the Operational Management Agreement. A Petroleum Fund secretariat has been established at the bank. The PRMA prescribes the qualifying instruments in which the funds should be invested, and mandates an Investment Advisory Committee to advice on the investment decisions and monitoring the performance of the funds (PRMA 2011).
- The bank also presents quarterly reports on the performance of the GPFs to the Minister of Finance and the Investment Advisory Committee;
- publishes semi-annual reports on the GPFs and submits those reports to Parliament;
- keeps books of accounts and records on the PHF and the GPFs;
- the Bank's Internal Audit Department audits books, records, systems, and procedures on the Petroleum Funds; and
- enters into contracts with the fund manager, specifying the services to be rendered, insurance and exposures, and accompanying fees.

Investment Advisory Committee (IAC)

The IAC is a seven-member committee of persons with proven competence in finance, investment, economics, business management, and related disciplines mandated to monitor the performance and management of the GPFs. The Minister for Finance and the governor of the Bank of Ghana nominate the members of the committee for appointment by the President. The IAC is responsible for the following tasks:

- advising the Minister and the Bank of Ghana on the investment strategy, instruments, guidelines, policy, and management strategy of the GPFs;
- formulating and proposing to the minister the investment policy for the GPFs;
- re-examining decisions made on the GPFs by the Minister in consultation with the governor in the absence of the advice of the IAC.

The Ministry of Finance

- publishes records of receipts, quarterly and annual reports of the management of Petroleum Funds;
- estimates and certifies the Benchmark revenue using a formula specified in the PRMA;
- develops the policy for the investment of the GPFs;
- is responsible for overall management of GPFs and transfers into and out of the Funds;
- in consultation with the Bank of Ghana, makes decisions on the investment strategy of the GPFs on the advice of the IAC;
- enters into Operational Management Agreements with the Bank of Ghana for the management of the GPFs;
- reviews the range of investment instruments designated as qualifying instruments;
- nominates members of the IAC in consultation with the governor for appointment by the President;
- gazettes and updates the gazette on the appointment of new members of the IAC; and
- determines the allowances paid to members of the IAC and PIAC.

The Auditor General

- is responsible for external audit of the Petroleum Funds;
- receives financial statements on the Petroleum Funds from the Minister and the Bank of Ghana and submits audited reports to Parliament;
- publishes audited reports of the Petroleum Funds.

Objectives

The Ghana Petroleum Funds (GPFs), a combination of the Ghana Stabilization Fund (GSF) and the Ghana Heritage Fund (GHF) are an important creation under the PRMA based on lessons, experiences, and global best practices. The GSF was established under Section 9 of the PRMA (Act 815) with the primary objective of cushioning the impact on or sustain public spending in periods of unexpected falls in petroleum revenues. After four years of operationalization, Section 12(5) of the Act 815 was amended by Act 893, allowing for withdrawals from GSF following shortfalls to also be transferred into the Contingency Fund and for debt repayment (Sinking Fund). The impact of this expanded purpose of withdrawals on the performance of the Fund will be evaluated subsequently in this section. The GHF, under Section 10, was established as an intergenerational equity fund to support development by future generations and to receive excess revenues.

The GSF was established under Section 9 of the PRMA (Act 815) with the primary objective of cushioning the impact on or sustain public spending in periods of unexpected falls in petroleum revenues. After four years of operationalization, Section 12(5) of the Act 815 was amended by Act 893, allowing for withdrawals from GSF following shortfalls to also be transferred into the Contingency Fund and for debt repayment (Sinking Fund). The impact of this expanded purpose of withdrawals on the performance of the Fund will be evaluated subsequently in this section. The GHF, under Section 10, was established as an intergenerational equityfund to support development by future generations and to receive excess revenues.

Investment and Risk Management Framework

This section evaluates the performance of the GPFs in terms of three subcategories:

i. Net realized income and returns;
ii. Overall growth and withdrawals (where applicable);
iii. Transparency and accountability.

Petroleum production started in the latter part of 2010 with revenue accruing in 2011. As mandated by the PRMA, 30% of revenue net of transfers to the NOC and allocation to the ABFA have always been transferred to the GPFs and, in some instances, exceptional transfers as required under the Act. Table 1 shows transfers into the GPFs since establishment.

Cumulatively, allocations to the GPFs reached US$1537.02 million at the end of December 2018, with the GSF contributing US$1,082.28 million, and the GHF, US$454.74 million. Figure 4 shows the realized returns net bank charges that have been earned annually on the funds since establishment (Bank of Ghana 2019). According to Section 12(2, 3, 4, and 5) of the Petroleum Revenue Management Act, the allowable amount to be withdrawn from the Ghana Stabilization Funds shall be the lesser of 25% of the estimated amount of the shortfall for the quarter or 75% of the balance standing to the credit of the Ghana Stabilization Fund of the financial year. An Investment Advisory Committee is charged with the responsibility of advising the Minister of Finance on the broad investment guidelines and overall management strategies relating to the Ghana Petroleum Funds. Section 30(1) and (2) provide further guidance on how the funds are to be invested. The Investment Advisory Committee (2020) lists the qualifying instruments as follows, consistent with Section 61 of the Petroleum Revenue Management Act (PRMA): overnight and call deposits, discount notes, treasury bills, short-term deposits, investment

Table 1 Annual allocations, withdrawals, and reserves of the GPFs

Year	*GSF (US$M)*	*GHF (US$M)*
2011	54.81	14.40
2012	16.88	7.24
2013	245.73	105.31
2014	271.76	116.47
2015	15.17	6.50
2016	29.51	12.65
2017	142.68	61.15
2018	305.72	131.02
Allocation Since Inception to Dec 2018	1082.28	454.74
Total Allocation Since Inception to Dec 2018	1537.02	
Withdrawals	714.61	–
Reserves (Dec 2018)	381.20	485.17
Total Reserves (Dec 2018)	866.38	

Source Bank of Ghana (2019)

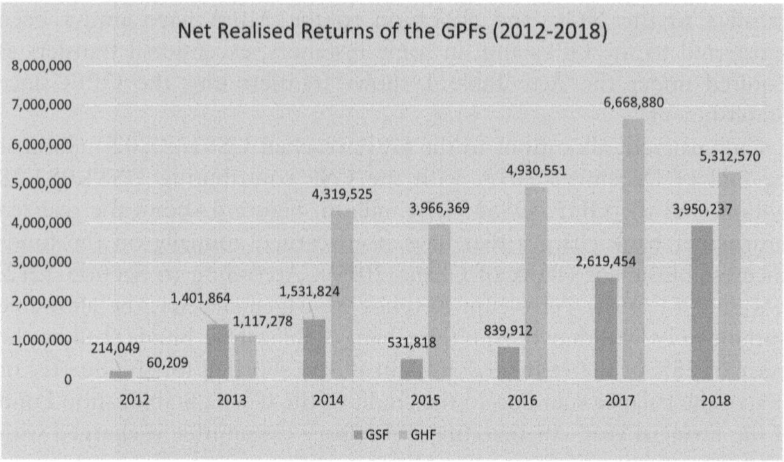

Fig. 4 Net Realized Returns of the GPFs in US Dollars (2012–2018) (*Source* Bank of Ghana [2019])

grade bonds, certificates of deposit, commercial papers, and medium-term notes.

Also the Bank of Ghana (2019) data reveals that the net returns largely depend on the size of the funds at any given time. Returns were relatively marginal in 2012 and started rising until the 2015 crude price meltdown, which affected petroleum revenues and transfers into the GPFs. For the same reason, the provision in law to withdraw from the GSF was triggered in that year and in 2016, bringing returns below 2012 levels as the fund was capped at US$150,000,000. The GHF, on the other hand, has grown steadily, translating into a corresponding growth in returns, falling slightly in 2015 compared to the previous year. The fund attained its highest returns since its establishment in 2017, but was reduced in 2018. The Bank of Ghana (2019) data further show that increases in the target range for the federal funds rate, lower longer-term inflation expectations, and lower term premiums in the US treasury market led to significant falls in the 10-year and 2-year treasury rates in 2018.

A review of the historical year-to-date returns on the GHF (see Fig. 5, Bank of Ghana [2019]) shows sharp movements in returns, negative returns in mid-2015, end-year 2016, and mid-2018, and the highest positive returns in mid-2016. The GSF returns are relatively stable, ranging

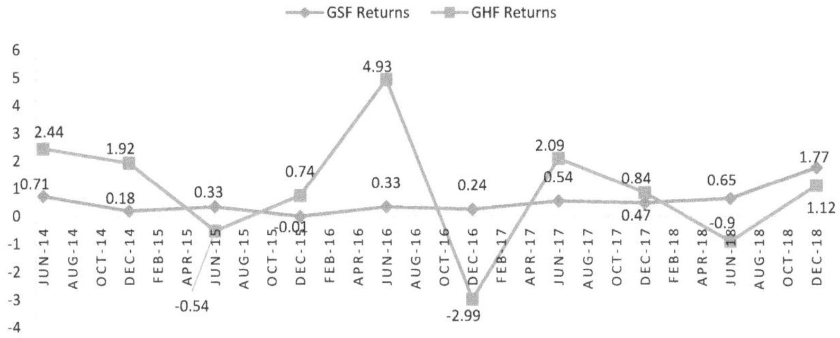

Fig. 5 Returns on the GPFs (*Source* Bank of Ghana 2019)

between −0.01% and 1.77%. The GSF returned a negative rate only in end-year 2015 and the highest positive return in end-year 2018. The movement in returns is largely reflective of the US treasury yield movements and the broader US, European, and other sovereign market dynamics. Returns are largely defined by the prescriptive nature of the investment portfolio in the PRMA.

Qualifying instruments as defined under the PRMA (Act 815) for the investment of the Ghana Petroleum Funds are prescribed as debt instruments denominated in internationally convertible currency, with interest or fixed amount equivalent to the interest of an investment grade security and issued or guaranteed by the International Monetary Fund (IMF), the World Bank, or by a sovereign State other than the Republic of Ghana. According to Section 61 (Interpretations, p. 32), a "qualifying instrument" is "a debt instrument denominated in international convertible currency that bears interest or a amount equivalent to interest, (i) that is of an investment grade security, (ii) that is issued by or guaranteed by the International Monetary Fund, World Bank, or by a sovereign State other than the Republic of Ghana, if the issuer or guarantor, has investment grade rating." This implies that the Petroleum funds cannot be invested locally.

The requirement for instruments backed by IMF and the World Bank reflects the nation's creditworthiness. Lending to sovereign states is usually considered a risk-free investment to a large extent as repayment obligations can be met with a number of fiscal options. However, returns on such investments are usually low to compensate for the low risks, as shown in Fig. 1.

Although the GPFs are affected by the prescriptive nature of the investment portfolio, withdrawals from the GSF, in particular, affect the reserves and hence its returns. Withdrawals from the GSF from inception to the end of 2018 and applications of funds withdrawn are shown in Table 2.

Technically, withdrawals have been made from the GSF on the basis of a shortfall in petroleum revenue in any quarter, in accordance with Section 12(2b), or the exercise of discretionary capping, in accordance with Section 23(3&4). Withdrawal from the GSF was first made in 2014 through the exercise of discretionary capping with another in 2015 following the fall in global crude prices and the resultant shortfall in the first quarter revenues for that year. Subsequently, different caps were imposed on the fund in the second half of 2015, 2016, and 2018.

The various destination accounts (Contingency Fund, Sinking Fund, and ABFA) into which transfers from the GSF have been made are shown in Table 2 (Bank of Ghana 2019). The Sinking Fund has been the most attractive destination, with about 87% of transfers going into it. This is contrary to the primary objective of the GSF and generally accepted principles of application of Stabilization Funds, but within the provisions of the PRMA as amended. The decision to cap the Stabilization Fund was part of government proposals submitted to Parliament during the process of amending the PRMA. This was to help the government sustainably service debt and interest payments. Therefore, instead of serving as a fiscal cushion during periods of low oil revenues, the fund is now largely a debt servicing strategy. In addition, the frequent transfers out of the fund in line with the caps do not allow for the growth of the Fund to serve its primary purpose. Total withdrawal from the GSF was US$714.1 million as at the end of 2018, nearly double the reserves for the same period. However, these withdrawals are allowed by the amended Act 893 (debt servicing), which deviates from the original intent of the Act 815 (fiscal stabilizer).

Process of Allocation and Spending into the Funds

Petroleum revenues as described under the Petroleum Revenue Management Act, 2011 (Act 815) as amended are made up of royalties, additional oil entitlements, surface rentals, corporate income taxes, capital gains taxes, proceeds from direct or indirect participation of government in

Table 2 Capping and withdrawal use of the GSF

Year	*Cap*	*Excess over cap*	*Contingency fund*	*Debt service/sinking fund*	*ABFA*	*Comment*
2014	250,000,000	H1 = 176,491,336.87 H2 = 129,193,351.53	H1 = 17,433,144	H1 = 159,058,193.47 H2 = 129,193,351.53		Capping
2015	H1 = 150,000,000	H1 = 53,685,578.98	–	–	53,685,579	1st quarter shortfall
	H2 = 173,755,072.85	H2 = 71,265,218.54	23,755,072.85	47,510,146		Capping
2016	200,000,000	–	–	–	–	Cap not attained
2017	–	–	–	–	–	No cap in H2
2018	300,000,000	H1 = 77,681,757.408 (Ex = 128,662,005.06) H2 = 206291095.81 (Ex = 81,203,231.15)		H1 = 77,681,757.408 H2 = 206,291,095.81	–	Capping
Total		714,608,340	41,188,217.00	619,734,544.22	53,685,579	
% of total withdrawal			5.76	86.72	7.51	

HI: First half of year; H2: Second half of year; Ex: Excess funds over cap
Source Bank of Ghana (2019)

petroleum operations, other payments by the National Oil Company such as corporate income tax, royalty, dividends, or any other amount, and other receipts from any petroleum operations, and the sale or export of petroleum.

The law stipulates that "payments into the Petroleum Holding Fund shall be net of the equity financing cost, including advances and interest of the carried and participating interests of the Republic; and the cash or the equivalent barrels of oil that shall be ceded to the national oil company out of the carried and participating interests recommended by the Minister and approved by Parliament" (PRMA 2011).

After deductions for the State's equity financing costs and transfers out of the carried and participating interest, the GPFs receive petroleum revenue from the PHF in excess of the Annual Budget Funding Amount (ABFA).

Transfers into the GPFs have largely been influenced by total petroleum revenue accrued for each year and the estimates of benchmark revenue for any financial year. Naturally, the more revenue accrues into the Petroleum Holding Fund (PHF), the more transfers into the GPFs in accordance with the allocation formula. However, this is not always the case, as transfers also depend on the estimated benchmark revenue. The benchmark revenue is an estimation of petroleum revenue expected by the government based on a seven-year moving average of crude oil prices and three-year average of government take of crude oil and natural gas.

Prior to the amendment of Act 815, a shortfall in actual revenue relative to the benchmark Annual Budget Funding Amount (ABFA) can lead to a withdrawal of up to 25% of the balance in the GSF or 75% of the quarter's shortfall with the possibility of doubling the withdrawal should the shortfall persist within the financial year. This meant that there could be quarters in which the GSF would not receive any transfers depending on the margin of shortfalls. However, under the amended Act (893), transfers into the GPFs are guaranteed given that even when petroleum revenues fall below the benchmark revenue, 30% of the actual revenues are transferred to the GPFs. Under Act 815, withdrawals could only be made for the purpose of alleviating shortfalls. The Minister of Finance is given the discretion to determine a cap for the GSF, beyond which subsequent transfers into the GSF may be applied for contingency or debt repayment purposes.

With the amendment, depending on the disparity between actual revenue (net transfers to the national oil company) and the benchmark revenue, withdrawals can be made in addition to the purpose of alleviating shortfalls, for contingency and debt repayment purposes. Since the benchmark revenue in most instances determines the allocation into the ABFA and GPFs and subsequently transfers from the GSF, the estimated benchmark revenue has been an important factor in the flow of funds into the GPFs. For instance, 2018 witnessed the highest allocation of US$305.72 for the GSF and US$131.02 for the GHF, although total revenue was slightly less than the 2014 levels as the benchmark revenue target was met and exceeded. In 2014, allocations into the GSF and GHF were US$271.76 and US$116.47, respectively (Bank of Ghana 2019).

Besides total petroleum revenue and the benchmark revenue estimates, another factor that affects flows into the GPFs is the allocation to the state-owned oil company for equity financing. A misinterpretation of the rules guiding transfers in the initial years of implementation of the PRMA affected transfers in 2012. Transfers into the GSF decreased drastically from US$54.81 million in 2011 to US$16.88 million in 2012, while the GHF decreased from US$14.40 million to US$7.24 million even as total petroleum revenues increased from US$444.13 million in 2011 to US$541.62 million in 2012. Given the margin of decline in transfers, the increased allocation to GNPC (from US$207.96 million in 2011 to US$230.95 million in 2012) alone could not have accounted for the reduction in transfers. This reduction could not also be attributed to the shortfall in revenue, as the shortfall was larger in 2011 (US$583.80) than in 2012 (US$ 260.50).

Transfers from the GHF can be made under two conditions:

- a parliamentary resolution by majority of members to transfer a portion of accrued interest at intervals of 15 years from the commencement of the Act; or
- upon consolidation of the GPFs into the Ghana Petroleum Wealth Fund, after which earnings from the wealth fund and dividends from the national oil company can be transferred into the ABFA.

As petroleum production is in its eleventh year, with more fields coming onstream, none of the above provisions has been triggered yet, because the oil resources have not depleted. The Act commenced in 2011, so the fifteen-year interval will not end until 2026. Also, there is no consensus in Parliament and nonstate actors on the need to further amend the law to allow withdrawal from the GHF.

Natural Resource Funds (NRFs) in Operation: Political Struggles over the Funds, Continuity, and Change

The essence of "Heritage" and "Stabilization" in the establishment of the GPFs has been questioned mostly by politicians, to whom investing the Heritage abroad and earning minimal interest while government borrows at exorbitant interest rates does not make economic sense. What constitutes "heritage," for instance, has been a subject of divergent public opinion. Expenditures on education and critical infrastructure such as railways, airports, major hospitals have often been cited as legacy spending with transgenerational benefits, which fall under the category of heritage.

The argument over how to use the Ghana Petroleum Funds polarized key sections of the Ghanaian society from political parties to civil society and policy organizations. Despite their differences, Ghana's dominant parties, the ruling New Patriotic Party (NPP) and the National Democratic Congress (NDC) believe in instituting measures to deal with oil price volatility and ensure intergenerational equity. A review of the manifestos of the two dominant parties for the 2016 elections shows the priority of fiscal governance of petroleum revenues. While the NPP pledged to enact a fiscal responsibility law to ensure prudent management of the Petroleum Funds, the NDC promised to accelerate the implementation of public financial management reforms such as the Integrated Financial Management Systems (GIFMIS). These notwithstanding, the post-election period generated debates on where and how the Ghana Petroleum Funds should be invested, due to the introduction of Free Senior High School (SHS) education policy and the public interest in how it would be funded.

According to the current Senior Minister, Mr. Yaw Osafo Marfo (NPP), ensuring intergenerational equitydoes not necessarily mean stashing money up and waiting for the future, as the future could be now (Nyabor 2017). He further reiterated that the state must undertake whatever measures are important to support the development of the country, especially for youth; hence, an amendment should be made to the PRMA

(2011) to allow a percentage of the heritage fund or the Petroleum Fund to be used to support the free secondary education policy: "The youth is the future, heritage is the youth, we (government) would make certain relevant amendments to make sure the economic benefits from the Petroleum Act" (Nyabor 2017).

This line of argument is premised on the reasoning that resources or petroleum wealth should be spent on current challenges to facilitate development into the future. Proponents of this idea believe that using the wealth now to create good education systems and youth employment, rather than investing in bonds, promises a better future for the country. In support of this proposal, the Minister for Planning argued that it is not reasonable to stash up money in bonds and stocks when the country is battling severe development deficits. As the then-ruling party, the NDC supported the use of the GHF to tackle economic challenges instead of saving it in financial instruments in the name of saving for the future. For instance, in 2014, the General Secretary of NDC, Mr. Johnson Asiedu Nketiah, argued that "in this current situation that we find ourselves in, it doesn't make economic sense to be keeping any money called Heritage Funds" (Obour 2014). Nketiah insisted that it made no economic sense for the Ghana to be saving the GHF which yields 1% interest while borrowing at 5 or 8% interest, and that if such situation continues, the country is going to leave behind a collapsed economy for future generations, thereby defeating the aim of the GHF (Obour 2014). However, as the opposition party since 2016, the NDC changed its viewpoint on the GHF from that of spend to save. For instance, during the debate about whether or not to use the GHF to finance the free SHS, the NDC insisted that doing so would compromise the fund. According to Deputy Minority Leader James Klutse Avedzi, it is not advisable to spend the GHF now, since it is an endowment for future generations (Ghanaweb 2017).

Civil society and policy organizations were also divided in their opinion about whether to save or spend the GHF. For instance, Dr. John Kwabena Kwakye of the Institute for Fiscal Studies (IFS) pushed for the investment of the GHF in infrastructural development. In his article "The Ghana Heritage Fund; To Save or Spend," he indicated that the opportunity cost of the GHF in its current form is undoubtedly high, as investments in human and physical capital that could promote growth would have to be forgone (Kwakye 2017). Although some Civil Society Organizations, such as IMANI Africa, did not dismiss the above line of argument outright, they did raise concerns about how the funds would be spent (Vaultz 2020). First, there is the sense that the government would not be fiscally disciplined in its spending and would not make judicious use of the funds for sustainable development. Second, the choice of which sector to invest to earn more benefits for future generations would be a matter of contention, since some people believe that investing in education should take priority over things such as road and transport infrastructure.

On the opposite side of the argument, Civil Society Organizations such as the Public Interest and Accountability Committee (PIAC) and the Africa Centre for Energy Policy (ACEP) believe that the principle underpinning the creation of a heritage fund is to provide for generations yet unborn, according to the strict reading of the PRMA (2011), which states: "The object of the Ghana Heritage Fund is to provide an endowment to support the development for future generations when petroleum reserves have depleted."

Section 10(4)PRMA, 2011 (Act 815) states: "Despite section 20, Parliament may by a resolution supported by the votes of a majority of members of Parliament at intervals of fifteen years from the date of commencement of this Act, review the restriction on transfers from the Ghana Heritage Fund and authorize a transfer of a portion of the accrued interest on the Ghana Heritage Fund into any other fund established by or under this Act." Section 20 of the PRMA further states: "Within one year after petroleum reserves are depleted, the monies held in both the Ghana Stabilization Fund and Ghana Heritage Fund shall be consolidated into a single Fund to be known as the Ghana Petroleum Wealth Fund after which the Ghana Stabilization Fund and the Ghana Heritage Fund shall cease to exist."

These give three potential channels for the use of the GHF. An amendment to these provisions in the PRMA, a resolution by majority of MPs in Parliament in 2026 or after the depletion of the petroleum resources, when the Stabilization and the Heritage Fund would be merged to form the Ghana Petroleum Wealth Fund. The earnings of the Ghana Petroleum Wealth Fund would be spent through the ABFA. Since none of these has happened, calls for the use of Heritage Fund will be mere talk.

The Natural Resource Governance Institute (NRGI) supports this and argues that "beyond potential economic challenges, the government could face legal challenges in accessing the GHF. The PRMA bars any withdrawals from the GHF in the short to-medium term. It only permits a partial withdrawal from interest accrued on the fund 15 years after the implementation of the PRMA—and then only with parliamentary approval. In practice, this means the government can only withdraw accrued interest from the fund from 2026 onward. Any attempt to withdraw from the GHF would require amendments to the PRMA" (Fusheini et al. 2017). This provision in the PRMA has not been reviewed and therefore the GHF cannot be used for government's programs or policies.

The debate over the use of the GHF was rekindled during the COVID-19 pandemic. On March 30, 2020, Finance Minister Ken Ofori-Atta presented an update on the economic impact of COVID-19 to Parliament, highlighting a projected revenue loss of GHS 5.6 billion, about 54% of projected loss of revenue, from oil, and non-oil revenue loss of GHS 2.254 billion. Additionally, the programmed initial intervention in the health sector to contain the spread of the disease required GHS 572 million. This, in addition to GHS 1250 million for the new Coronavirus Alleviation Programme (CAP) to support businesses and households, widens the fiscal gap for the financial year, hence, yielding a total fiscal unplanned deficit of GHS 9.505 billion. In order to fund this fiscal deficit, which could be more than the 5% allowed by Ghana's fiscal responsibility law, the Minister made a number of proposals such as reducing the cap on the Stabilization Fund, reducing expenditure, decreasing allocation to the National oil company and an amendment to the PRMA to use a portion of the GHF to support the budget (Dapaah 2020). Civil Society Organizations including PIAC, ACEP, NRGI, Friends of the Nation, among others issued statements to oppose any potential amendments that would allow use of the GHF to support the fight against COVID-19. As PIAC noted, the original intent of the establishment of the GHF must be adhered to:

> It is worth noting that there have been previous attempts by successive governments to access the Heritage Fund for various reasons, to which PIAC objected. The Committee's position is consistent with its understanding of the spirit of the PRMA on the management and use of the Ghana Heritage Fund (GHF) – to cater for the intergenerational ownership of the petroleum resource as well as the shared compensation on the adverse effects of petroleum extraction. (Dapaah 2020)

These civil society groups' opposition also aligns with the position of the NDC, the opposition party in Parliament (Emmanuel 2020). Faced with these oppositions, the Finance Minister appeared to have rescinded, indicating a few days later that his proposal was meant just to trigger a debate (Agyeman 2020).

The operation of the other Ghanaian oil fund has also generated some debate, although not at as much of an emotive and controversial level as the GHF. emotional level as the GHF. For instance, in line with the provisions of the PRMA (2011), the GSF was meant to provide fiscal buffer or budgetary support in periods of shortfalls in petroleum revenues. However, this fund has been amended over the years, capping it and allowing for the excess revenues from it to be used in setting up a sinking fund to pay off Ghana's debts. It was initially capped at $250 million in 2014 and reduced to $100 million in 2016, and has stood at $300 million since 2017 (Gyeyir 2019). This development has generated sharp criticism from Ghana's energy watch organizations, and scholars such as Dr. John Gatsi of the University of Cape Coast Business School, who argued that although amortization of loans and the sinking fund were all intended to generate a diversified fiscal buffer for the economy, it creates opportunities for excessive discretionary powers and the concomitant challenge of how to checks and restrict such discretion (Ackah et al. 2020).

In fact, according to organizations such as PIAC and other institutions such as ACEP, capping the fund and using the money to settle debt is not the best economic decision, given the purpose of the fund. These organizations insist that such uses are clear deviations from the original intent of the GSF as a bulwark against volatility in the medium to long term (ACEP 2015). They also argued that capping the fund has afforded the government the luxury of using the excess funds to pay unsustainable debts. Reiterating their position, the ACEP explained that capping the fund puts the country in a situation in which the government has unfettered liberty to contract loans, use them for other purposes which cannot be tracked or monitored, and use oil funds to repay those loans. ACEP, conversely, proposes that petroleum wealth be spent on two sectors, such as agriculture and education, for the first three years, and move on to develop two other vital sectors of the economy for the next three years. In doing so, agriculture and education, for instance, would receive oil revenues only as complements and not as substitutes for what the government already gives those two sectors through the national budget. In addition to the above critical voices within Ghana, the NRGI has also expressed the concern that the GSF was no longer merely being used for stabilizing oil revenues (Fusheini et al. 2017). In this view, the capping of and transfers to the sinking fund have been used to secure additional borrowing to ease government debt financing challenges. As a result, the low-interest savings of the GSF are not unreasonable, since the government has been engaging in very high-interest borrowing over the years, which could not simply be offset by low-interest savings. Indeed, NRGI cautioned that, international experience has shown that continuously altering the legislative framework for managing Petroleum Funds leads to poor outcomes (Fusheini et al. 2017).

Another area of contention regarding Ghana's oil funds is the issue of investment returns and oversight. Returns on the GSF and GHF have averaged 0.521% and 0.965%, respectively, over the period between 2014 and 2018. Citizens have expressed concerns about the low return on investment of the GPFs vis-à-vis the interest on government borrowed funds, with a 10-year bond in 2018 priced at 7.625% and described by the Ministry of Finance as the lowest interest rate ever (Ministry of Finance 2018). The PRMA prescribes the qualifying instruments for investment of the GPFs. Many have, however, called for a review of the instruments to grant some flexibility to the Bank of Ghana and the Investment Advisory Committee, to vary the investment portfolio, and allow for reasonable risks and rewards (PIAC 2018).

Section 23(3) of Act 815 provides that the accumulated resources in the GSF shall not exceed an amount recommended by the Minister and approved by Parliament, and that the amount be reviewed from time to time as necessitated by macroeconomic conditions. The cap has varied widely between 2014 and 2018, ranging between US$150 million and US$300 million. It is difficult to fathom what other economic conditions besides the oil price drop of 2015–16 might have necessitated the various caps imposed on the GSF. Even with the aforementioned price drop, disclosure of the economic parameters that informed the various caps was still important.

In accordance with Section 23(4), once the amount set as the cap for the GSF is attained, subsequent transfers into the Fund are allocated as transfers into the Contingency Fund or Sinking Fund (debt repayment) as approved by Parliament. About 7.5% of over US$714 million has been applied to shore up shortfalls in the budget, the bulk (86.7%) of which was transferred into the Sinking Fund. The Minister of Finance is required by Section 48 of Act 815 to submit to Parliament and publish an annual report on the Petroleum Funds. Whereas Section 48(2)(a)(ii) requires the Minister to publish withdrawals from the GSF and GHF, Section 48(2)(f) in particular mandates the publication of liabilities of government borrowings by the Minister. Given the amount that has been transferred into the Sinking Fund, these publications ought to include details of the debts that are repaid with proceeds from the Sinking Fund.

This chapter argues that, given the IAC's role as outlined, the prescriptions under Section 61 of Act 815 renders futile the potency of the IAC to consider a broad range of instruments to maximize returns. A key consideration in the formulation of an investment policy, one of the functions of the IAC, is a determination of the choice of instruments or combination of instruments to minimize exposure and maximize returns. The level of risks has, however, been determined by the definition of the qualifying instruments. This chapter argues that the advisory role of the IAC makes it ineffective as the Minister is not bound by its advice and decisions. It is important to state that Section 38 of Act 815 attempts to remedy situations in which the Minister may make decisions without the advice of the IAC. It allows the Minister, in urgent circumstances, to make decisions in consultation with the governor of the Central Bank, but also makes provisions for the IAC to provide its advice post factum. Since the Minister of Finance is faced with a wide range of competing and sometimes conflicting economic and socioeconomic considerations, how practical would it be to make the decision of a Committee binding?

A final challenge in the operation of Ghana's oil funds is transparency. To be sure, transparency and reporting by various entities are integral parts of the framework for the management of petroleum revenues. Indeed, the Bank of Ghana, the Ministry of Finance, and the Public Interest and Accountability Committee (PIAC) have been producing quarterly, half-year, and annual reports on the management of the GPFs and the larger petroleum revenues. However, only the 2015 and 2016 reports of the Auditor General on the financial statements of the Petroleum Funds by the Bank of Ghana are published on the Ghana Audit Service's website. These reports are expected to be submitted to Parliament by June at the latest and published by July of the following year. The advice provided by the Investment Advisory Committee (IAC) and reports of the Auditor General are often not published as part of the Minister's annual report to Parliament per the requirements of Section 48 of Act 815. Real-time data on the value and investment instruments of

GSF and GHF are not provided on the website of the Bank of Ghana, and fund managers (equity and external) are not publicly disclosed.

Summary, Conclusion, and Recommendations

Upon the commercial discovery of petroleum in 2007, Ghana established a framework to guide the use of proceeds in an attempt to leverage these resources to further the country's development while making provisions for savings and investment. The Ghana Petroleum Funds (GPFs) were included in this framework in recognition of the entitlement of future generations to this shared resource and the need for savings and investment to mitigate the volatile nature of revenues and other negative macroeconomic effects of large revenue flows.

This framework has resulted in the GPFs benefiting from allocations of some US$1537.02 million to the Funds, with the GSF receiving US$1082.28 million and the GHF receiving US$454.74 million. Of the amount allocated to the GSF, only 35% (US$381.20) remains in the Fund, with the remainder (US$714.61) withdrawn in accordance with provisions of Act 815. The Funds were largely applied for the payment of statutory debts, with 7.5% of withdrawals used for the primary purpose of stabilizing shortfalls in the budget.

Returns on the GPFs have been generally low, averaging 0.743% between 2012 and 2018, a reflection of the low-risk portfolio of the Funds as mandated by Act 815. One reason for this has been the limited instruments prescribed by the law. In addition, the Investment Advisory Committee (IAC), which is mandated by law to develop investment policy and advise on the strategy and choice of instruments for the investment and management of the GPFs, was not active between 2016 and 2018, which prevented the Funds from benefitting from the technical advice of the Committee. We recommend that as the Ghana Petroleum Funds grow, considering an amendment of the PRMA to reduce the rigidities of the investment instruments and provide a greater scope for the Investment Advisory Committee to propose a wider portfolio of investment options to enhance returns while retaining value of the funds would also be useful. For the Stabilization Fund to perform its intended role effectively, withdrawals ought to be based on publicly disclosed sound economic rationale with less arbitrariness. As the GPFs grow, considering an amendment of the PRMA to reduce the rigidities of the investment instruments and provide a greater scope for the IAC to propose a wider portfolio of investment options to enhance returns while retaining value of the funds would also be useful. Transparency on the management of the funds should be improved with the provision of real-time information and data on the asset allocations, performance, and management of the GPFs.

Perhaps by far, the controversies surrounding Ghana's oil funds have largely revolved around the GHF, specifically questions over whether to draw on the Heritage Fund to accomplish government's plans and programs or whether to save the money for future generations as was originally intended. Our position on this debate is that the use of the Heritage Fund now may not be possible for at least two main reasons. First, policymakers have access to more than 90% of the oil revenues for capital, goods and services, and funding of the national oil company. While transparency has improved over time, accountability has become a challenge in some cases. Almost every year, PIAC or ACEP finds potential cases of noncompliance with the PRMA (Public Interest and Accountability Committee 2011–2018). Unfortunately, nothing happens after from sensational headlines by newspapers. So, if more than 90% cannot be properly accounted for, why should the remaining in the form of GHF be added. So, trust has become a central theme in these debates. Second, hard lesson has been learnt from Ghana's experiences in the management of mineral revenues in the way of saving for tomorrow and this historical experience will continue to shape debate about the use of the Fund.

References

Ablo A, Asamoah VK (2018) Local participation, institutions and land acquisition for energy infrastructure: The case of the Atuabo gas project in Ghana. Energy Research and Social Science 41:191–198

Abraham K (2019) Petroleum revenue management in Ghana: The epoch of high expectation in perspective. Journal of Sustainable Development Law and Policy 10 (1):32–55

Ackah I, Bobio C, Graham E, Oppong CK (2020) Balancing debt with sustainability? Fiscal policy and the future of petroleum revenue management in Ghana. Energy Research and Social Science 67. https://doi.org/10.1016/j.erss.2020.101516

Africa Centre for Energy Policy (2015) Review of the Petroleum Revenue Management Act 2011 (Act 815) http://acepghana.com/wp-content/uploads/2013/12/ATT00051.pdf

Agbefu R (2011) The discovery of oil in Ghana: Meeting the expectations of local people. Mphil thesis, University of Oslo

Agyeman N (2020) Finance Minister clears confusion over proposal to use Heritage Fund in wake of COVID-19. Graphic, 5 Apr. https://www.graphic.com.gh/news/general-news/finance-minister-clears-confusion-over-proposal-to-use-heritage-fund-in-wake-of-covid-19.html

Alagidede P, Akpoza A (2012) Sovereign wealth funds and oil discovery: Lessons for Ghana, an emerging oil exporter. AfricaGrowth Agenda Oct/Dec:7–10

Andrews N (2013) Community expectations from Ghana's new oil find: conceptualizing corporate social responsibility as a grassroots-oriented process. Africa Today 60(1):55–75

Aryeetey E, Ackah I (2018) The boom, the bust, and the dynamics of oil resource management in Ghana. WIDER Working Paper No. 2018/89

Bank of Ghana. Semi-annual Reports on the Ghana Petroleum Funds. https://www.bog.gov.gh/about-the-bank/public-notices?start=30

Bank of Ghana (2019) Petroleum Holding & Ghana Petroleum Funds Semi Annual Report: Jan 2–Jun 28. https://www.bog.gov.gh/wp-content/uploads/2019/08/Semi-Annual-Report-H1-2019.pdf

Bawumia M, Halland H (2017) Oil discovery and macroeconomic management. Extractive Industries 220

BBC News (2007) Ghana 'will be an African tiger'. 19 June. https://news.bbc.co.uk/2/hi/africa/6766527.stm

Dapaah E (2020) PIAC kicks against usage of Heritage Fund for COVID-19 fight. Citinewsroom. 6 Apr. https://citinewsroom.com/2020/04/piac-kicks-against-usage-of-heritage-fund-for-covid-19-fight/

Emmanuel K (2020) NDC cautions gov't against use of Heritage Fund to tackle COVID-19. Pulse, 3 Apr. https://www.pulse.com.gh/news/politics/ndc-cautions-govt-against-use-of-heritage-fund-to-tackle-covid-19/2snx113

Fusheini A et al. (2017) Funding Ghana's 'Free' Senior High School with oil revenue: Sober reection required. Briefing. March. https://resourcegovernance.org/sites/default/files/documents/funding_ghanas_free_senior_high_school_with_oil_revenue_0.pdf

Ghanaweb (2017) It's wrong to use Heritage Fund for free SHS – James Avedzi. 15 Feb. https://www.ghanaweb.com/GhanaHomePage/NewsArchive/It-s-wrong-to-use-Heritage-Fund-for-free-SHS-James-Avedzi-510320

Graham E, Ackah I, Gyampo RE (2016) Politics of oil and gas in Ghana. Insight on Africa 8(2):131–141

Gyampo R (2010) Saving Ghana from its oil: A critical assessment of preparations so far made. African Research Review 4(3):1–16

Gyeyir DM (2019) The Ghana Stabilization Fund: Relevance and impact so far. Energy Policy 135:110989.

Investment Advisory Committee (2020) The fund's growth. https://www.iacghana.com
Kpodo K (2010) Ghana chiefs in oil region demand 10% of revenues. Reuters, 17 Nove. https://af.reuters.com/article/idAFJOE6AH02Z20101118
Kwakye JK (2017) The Ghana Heritage Fund: To save or to spend? Graphic Online, 16 March. https://www.graphic.com.gh/features/opinion/the-ghana-heritage-fund-to-save-or-to-spend.html
Ministry of Finance (2018). Ghana raises US$2.0 billion in Eurobonds with the lowest interest rate ever. https://www.mofep.gov.gh/press-release/2018-05-14/ghana-raises-US%242.0-billion-in-eurobonds-with-the-lowest-interest-rate-ever
Nyabor J (2017) Gov't to fund free SHS with Heritage Fund—Osafo Maafo. CitiFMonline,15 February. https://www.modernghana.com/news/755753/govt-to-fund-free-shs-with-heritage-fund-osafo.html
Obour S (2014) Government should spend Heritage Fund if... - Franklin Cudjoe. Daily Graphic, 6 May. https://www.modernghana.com/news/539336/government-should-spend-heritage-fund-if-frank.html
Okpanachi E, Andrews N (2012) Preventing the oil "resource curse" in Ghana: Lessons from Nigeria. World Futures 68(6):430–450
Osei-Tutu J (2012) Managing expectations and tensions in Ghana's oil-rich Western Region. SAIIA Policy Briefing 55. August. https://www.files.ethz.ch/isn/152834/saia_spb_55_osei-tutu_20120828.pdf
Panford K (2014) An exploratory survey of petroleum skills and training in Ghana. Africa Today 60(3):57–80
Petroleum Revenue Management Act (2011) Act 815 as amended by the Petroleum Revenue Management Act, 2015 (Act 893)
Public Interest and Accountability Committee (2011–2018) Reports on the management of petroleum revenues in Ghana. http://www.piacghana.org/portal/5/25/piac-reports
Public Interest and Accountability Committee (2018) Why the absence of the investment advisory committee is a significant violation of the PRMA. https://www.piacghana.org/portal/12/13/257/why-the-absence-of-the-investment-advisory-committee-is-a-significant-violation-of-the-prma?fbclid=IwAR0uqpGxNbYMu04PYdzj-eAb8FIRvXfzMKd3tZ6vWUAlRtUUzExot1vI388
Public Interest and Accountability Committee Reports of District Engagements and ABFA Project Inspections. http://www.piacghana.org/portal/5/26/secretariat-reports
Standing A (2014) Ghana's extractive industries and community benefit sharing: The case for cash transfers. Resources Policy, 40:74–82
Tsikata F (1977) The vicissitudes of mineral policy in Ghana. Resources Policy 23(1–2): 9–14

Vaultz (2020) Suspend free SHS, NABCo, and new register -IMANI advises govt. Vaults Daily Brief, April 7. https://thevaultzmag.com/index.php/vaultz-daily-brief/suspend-free-shs-nabco-and-new-register-imani-advises-govt

To Save or Not to Save: State Governments and the Construction of Privilege Over the Creation of Oil Funds in Nigeria

Eyene Okpanachi

Introduction

Nigeria is the largest oil producer in Africa, producing about 2 million barrels per day (bpd) of oil in 2019 (International Monetary Fund 2019), even though the country has been unable to meet its peak production volume of about 2.5 bpd as a result of violence, including sabotage and theft, in the Niger Delta, the oil producing region, and, more recently, OPEC oil cuts (US Energy Information Administration 2020). The country has about 36,890 billion barrels of crude oil reserves in 2019 (OPEC 2020), and an estimated 200.4 trillion cubic feet (tcf) of proved

The original version of this chapter was revised: In Chapter 'Privilege' in the title was spelt as 'Priviledge' correction has been updated. The correction to this chapter is available at https://doi.org/10.1007/978-3-030-78251-1_12

E. Okpanachi (✉)
University of South Wales, Pontypridd, UK
e-mail: eyene.okpanachi@southwales.ac.uk; eyeneokpanachi@uvic.ca

University of Victoria, Victoria, BC, Canada

E. Okpanachi and R. Chowdhari Tremblay (eds.), *The Political Economy of Natural Resource Funds*, International Political Economy Series,
https://doi.org/10.1007/978-3-030-78251-1_7

natural gas reserves in 2019 which is the largest on the African continent (Oil & Gas Journal cited in U.S. Energy Information Administration 2020). Despite some improvement in non-oil tax revenues collection in Nigeria in recent years, oil and gas still dominate Nigeria's revenues profile, accounting for 86% of Nigeria's total exports revenues (OPEC 2020), with non-oil revenues representing a mere 3.6% of GDP by 2019 (International Monetary Fund 2019).

Oil and natural gas resources are found in various quantities in ten of the 36 states, where most of the country's 606 oil fields, consisting of 355 onshore and 251 offshore fields, are located (NNPC 2010). About 91.5% of oil production takes place in six of the 10 oil-producing states located in the Niger Delta or South–South geopolitical zone. Together, these states—Akwa Ibom, Bayelsa, Cross Rivers, Delta, Edo, and Rivers—account for 15% of Nigeria's total population. Of these Niger Delta states, four—Akwa Ibom, Bayelsa, Rivers, and Delta—are regarded as the "core" oil-producing areas, because most of the oil production takes place in these states. Edo and Cross Rivers states in the South–South geopolitical zone, Ondo and Lagos states in the southwest zone, and Abia and Imo in the southeast zone, are the "non-core" oil producing states, accounting for 8.5% of total oil production (Iledare and Suberu 2012). While oil production commenced in Lagos in southern Nigeria in 2016, making the state to become the newest oil-producing state in the country, exploration outside the Niger Delta has not led to oil production. This outcome is even more disappointing outside southern Nigeria, particularly in the northern part of the country where significant oil explorations have taken place, propelled by successive governments' desire to diversify Nigeria's oil resource base, enhance energy security, and increase revenues.[1] While crude oil has been discovered in some sedimentary basin in the north, particularly the Chad basin and Upper Benue Trough of the Gongola Basin, production has yet to commence, not least as a result of insecurity arising from the activities of the Islamist terrorist group known as Boko Haram operating in the northeast region. In July 2017, three members of an oil exploration team were kidnapped by Boko Haram, and it was not until February 2018 that the kidnapped persons were released (BBC News 2018).

[1] Exploration of oil in the north is also the Nigerian governments' attempt to pacify the region's elites' clamour for states in the region which they believe are richly endowed with oil reserves to become "oil-producing" states and benefit from additional revenues from oil like the states in the Niger Delta.

The 1999 Nigerian Constitution provided for the sharing of oil revenues based on a formula designed by the Revenue Mobilization Allocation and Fiscal Commission (RMAFC), but with the approval of the National Assembly. The constitution provided that all revenues (including oil revenues), apart from income taxes of military men, residents of the federal capital territory (FCT) Abuja, and expatriates, should be deposited in a fund known as the Federation Account,[2] from which money should be shared between the states, local governments, and the federal government, and between states and local governments themselves, based on the formula designed by the RMAFC. Section 62(2) of the 1999 Constitution empowered the President to submit proposals for revenue allocation from the Federation Account to the National Assembly upon receipt of this proposal from the RMAFC. In determining the formula for revenue allocation presented by the President, the 1999 Constitution further provided that the National Assembly should be guided by certain allocation principles, "especially those of population, equality of States, internal revenue generation, land mass, terrain as well as population density" (Federal Republic of Nigeria 1999). However, there was a proviso that in taking account of these principles, the National Assembly must ensure that the "principle of derivation shall be constantly reflected in any approved formula as being not less than thirteen percent of the revenues accruing to the Federation Account directly from any natural resources" (Federal Republic of Nigeria 1999, p. 66). In other words, although oil revenues are collected by the federal government, the 1999 constitution provided for the sharing of 13% of the gross oil revenue among the oil-producing states in proportion to the volume of oil/gas produced in each of these states. The remaining revenues are paid directly into the Federation Account and distributed among the three levels of government as follows: the federal government receives 52.68%; the state governments, 26.72%; and local governments, 20.60%.[3]

[2] It should be reiterated that oil, which accounts for almost 70% of total federally collected revenue, is the major source of revenue for the Federation Account. Other sources of income include VAT, custom and excise duties, and corporate income tax.

[3] The horizontal sharing of revenues in the Federation Account is calculated using the following allocation formula: equality 40%; population 30%; land mass and terrain 10%; internally generated revenue 10%, and social development(education enrolment, health, and water) 10%.

The Political Economy of Nigeria's Oil Dependency and the Challenges of Managing Oil Wealth

Oil has not always played a dominant role in the Nigerian economy. Nigeria had traditionally been dependent on agricultural commodities such as palm oil, groundnuts, cocoa, and, to some extent, a few minerals such as coal, columbite, and tin. All this began to change after the end of the civil war in 1970, and changed even more rapidly following the 1973–1974 OPEC-induced crisis, which ushered in massive revenues accruing from oil, made Nigeria to become chronically dependent oil, a monolithic resource base that is subject to the vagaries of international market prices, and changed attitudes toward the development of the non-oil sectors on which Nigeria had depended before the oil boom. For instance, after the end of the civil war in 1970 and, especially, during the 1973–1974 OPEC-induced oil crisis, Nigeria's crude oil production volume not only increased, there was also a significant rise in oil windfalls accruing to the country. Between 1970 and 1983, the Nigerian state earned $140 billion from oil (Watts 1992). The contribution of the oil sector to Nigeria's foreign exchange income increased from 8.5% in 1964 to 86% in 1974 (Madujibeya 1976, p. 290). In addition, the percentage of oil revenues—royalties, profit tax, domestic crude sales, and others—rose from 26% of federally collected revenues in 1970 to 81% in 1980 (Iledare and Suberu 2012, p. 229). Given this massive dependence on oil revenues, the economy collapsed with the end of the oil boom (Watts 1992, p. 27), as oil exports fell to 708,000 bpd in 1981 from a high of 2.2 million bpd (Onoh 1983, p. 91). The oil price boom-and-bust also saw a dramatic change in Nigeria's oil revenue: revenues from oil exports grew from $718 million in 1970 to $25 billion in 1978, and then declined, with the collapse of oil prices, to $4.7 billion in 1986 (Iledare and Suberu 2012, p. 229).

To be sure, oil windfalls did help Nigeria to carry out economic modernization. Watts (1992, p. 27) noted, for instance, that there was an almost 40% yearly growth in government revenues during the booms of the 1970s, and these enormous revenues propelled massive state investment in the economy and some forms of industrialization, which translated into almost a tripling of the national manufacturing index between 1972 and 1980 (Watts 1992, p. 27). However, Brian Pinto (1987) has also drawn attention to the policy choices during the oil boom that made adjustment to the bust period difficult and undermined the saving of windfalls that accrued to the country during the boom. A major way this took place, Pinto (1987) noted, was through spending on unsustainable public projects while the agricultural sector, which was the major source of revenue outside of the oil industry and the major source of employment compared to the limited opportunities in the oil sector, was neglected. Overreliance on oil exports and lack of investment in the non-oil sectors of the economy therefore led to the emergence of Dutch disease in the economy marked by a decline in agriculture and ballooning inflation, current account deficits, and reckless borrowing (Pinto 1987, p. 419) following the collapse of oil prices in 1982, with the external debt rising steeply to $18 billion, or 160% of exports, by 1984 (Lewis 2007, p. 161). In addition, oil windfalls provided opportunities for massive state intervention in the economy for political rather than economic purposes. A good example of this, as noted by Peter Lewis, was the "proliferation of public enterprises" driven not by economic efficiency, but by the need to expand rents and patronage "as part of an expanding domain of state-supported entitlements that bolstered the political reach of military elites and their civilian allies" (Lewis 2007, p. 143). For instance, where the number of state enterprises stood at 250 before the boom in 1970, it increased to over 800 a decade later (Lewis 2007, p. 142).

Responding to Oil Price Volatility through the Creation of an Oil Fund—The Excess Crude Account (ECA)

In order to avert the crises of the 1970s, the Olusegun Obasanjo government created a fiscal buffer in the form of an entirely new account, known as the Excess Crude Account (ECA), in 2004, with the aim of saving excess oil receipts above the budgeted oil price. The ECA's primary objective was to insulate the Nigerian economy from the volatility of the global oil market by delinking government expenditures from oil revenues (CBN 2012, p. 3). For example, if the benchmark price for a barrel of oil in a given year's Appropriation Act is $50 and oil price increases to $87 per barrel, the amount beyond the budget benchmark price (in this case, $37 per barrel) will be deposited in the ECA. However, what this means is that when oil prices fall below $50 per barrel, the ECA will be starved of funds.

It should also be pointed out that the establishment of the ECA was part of the broader economic reform policies instituted by President Obasanjo, especially during his second term in office (2003–2007). Even though the government described the economic reform policy known as the National Economic Empowerment Development Strategy (NEEDS), of which the creation of ECA was a major component, as "home-grown," the policy was also a response to the pressures of the global political economy. Indeed, the ECA was created after a meeting of the Honorary International Investors Council (HIIC), an organization of prominent investors from around the world that advises the Nigerian government on matters pertaining to the country's economic development (Obiejesi 2018). President Obasanjo himself noted the foreign impetus of the economic reforms when he stated: "Our home-grown reform programme, encapsulated in National Economic Empowerment and Development Strategy (NEEDS), was monitored on a quarterly basis by the IMF, at the invitation of the government, in order to provide the global community with objective assessments and reports on how well we were doing" (Obasanjo 2005). Against this background, it is therefore not surprising that a major purpose to which funds drawn from the ECA were put, once the rise in oil prices resulted in accumulation in the account, was the servicing of Nigeria's external debt obligation. Specifically, about $12.4bn was withdrawn from the account to offset Nigeria's debt to the Paris Club (Alli 2012) in a process supervised by the IMF. The payment of the debt, facilitated by savings in the ECA, enabled Nigeria to get $18 billion written off its overall debt by the creditor nations.

A very important note at this point is that Obasanjo created this account single-handedly, without the consent of the governors or the National Assembly. Despite its usefulness as an important mechanism for saving oil windfalls, the account's questionable foundational legitimacy made it the focus of controversy, with state governors and the Chairman of the Revenue Mobilization, Allocation and Fiscal Commission (RMAFC), the agency that monitors the accruals to and disbursement of revenue from the Federation Account, and federal legislators declaring the ECA illegal (Murray and Akpe 2006). As will be discussed below, the institutional rule regarding revenue sharing in Nigeria, which stipulates that all revenues should be credited to the Federation Account and shared by all governments, is central to this controversy.

Political Opposition to the Excess Crude Account and the Creation of the Sovereign Wealth Fund

Realizing that they could not force President Obasanjo to close the ECA, the governors mounted pressure to share the money in the account, which they believed should be deposited in the Federation Account and shared on a monthly basis. However, this pressure did not achieve much, as Obasanjo used the money in the ECA to make payments for activities that were considered the exclusive responsibility of the federal government. This decision provoked an outcry from the governors, who accused the President of short-changing them by using revenues meant for the three governments of the federation to pay for the federal government's own assignments. With the installation of the Umaru Yar'adua government in 2007, the governors grew bolder and more powerful in their pressure on the President to share money from the ECA. They took the federal government to court and eventually received concessions from President Yar'adua, who distributed the account periodically to all governments, so much so that the account was drawn down from $20 billion in 2007 to less than $500 million in 2010 (Onuah 2010b). At the time the Jonathan government, which had experienced some sustained oil booms, handed over power in 2015, what was left in the ECA was a paltry $2.3 billion. This number carried the implication that the economy had lacked a buffer as the oil price collapse of 2014 continued, leading to deferred investments in human capital and infrastructures by federal, state, and local governments, and also to the inability of state governments to meet even basic expenditures such as teacher salaries (Kay and Wallace 2016). To be sure, the ECA did stabilize the economy, as was the case during the 2008–2010 global economic crisis, when funds from the account helped Nigeria navigate the crisis without seeking bailout from the IMF/World Bank as some other countries did. However, a number of unaccounted withdrawals were also made from the ECA, which led to the dwindling of the account even during periods of increasing oil prices (NEITI 2017, p. 170). Even though these withdrawals were carried out by the federal government itself, pressures from state governors for the sharing of funds in the excess crude account contributed largely to the ECA's depletion.

The point to note is that the ECA has led to countless litigations, wars of words, allegations, and denials between the federal and state governments. States have used the court case they instituted against the constitutionality of the ECA as a bargaining chip to gain proceeds in the account shared through an out-of-court settlement, for which the federal government has often pleaded and which the judiciary has always granted since the governors brought the case before the Supreme Court in 2008. These pressures by state governments have led to significant drawdown of the ECA, even though President Obasanjo, unlike the two presidents after him, was able to wade off some of these pressures to share from the account.

State governors exploited the deficiencies of the ECA to push for its funds to be shared, which nearly depleted the account. This led to serious consideration by the federal government about how to prevent such a situation in the future. To reiterate, a major problem with the ECA was that it was unilaterally established as an exigent administrative arrangement, with little consideration given to the institutionalization of accompanying legal rules regarding its legitimacy and consequent operation, including rules regarding the withdrawal and allocation of funds from the ECA, and the specific authority responsible for approving withdrawals (NEITI 2017). Because the ECA held money that constitutionally belongs to the three levels of government—the federal, state, and local governments—the institutional weakness regarding its creation and operation opened it up to incessant pressures from state governors who insisted that the ECA should not exist in the first place and that the money in the account should be shared between the governments. With no clear legal framework to guide the distribution of the money in the ECA, it became a ready source for extra budgetary spending, and pressures for the sharing of its accruals led to its depletion. According to NEITI, "between 2005 and 2015, the sum of $201.2 billion accrued to the Excess Crude Account, but $204.7 billion was withdrawn from the same account. In other words, outflows were 102% of inflows" (NEITI 2017).

To guarantee sustained savings of Nigeria's oil windfall, which seemed not to have worked well with the ECA, the Goodluck Jonathan government made efforts to rectify and transcend the institutional weakness of the ECA by forwarding an executive bill to the National Assembly, following an earlier National Economic Council (NEC),[4] aimed at instituting "the legal underpinning for national savings" through the creation of a national sovereign fund (Shehu 2010) to take over the ECA and manage the Sovereign Wealth Fund (SWF). However, generating the political consensus for the creation of a SWF proved difficult, as state governors also opposed the move, with the governors instituting a case at the Supreme Court in order to stop it. After months of disagreement, a consensus appeared to have been reached when the President's bill for the creation of the Nigeria Sovereign Investment Authority (NSIA) was passed by both chambers of the National Assembly in May 2011 and assented to by President Jonathan thereafter, even though the court case instituted by the governors was still subsisting.

Nigeria's Sovereign Wealth Fund: The Nigeria Sovereign Investment Authority (NSIA)

The NSIA was established by the *Nigeria Sovereign Investment Authority Act* (the "Act"), which also specifies its ownership, powers, legal characteristics, and policy objectives. According to the Act, the NSIA is owned by the governments of the federation—the federal government, the state governments, the local governments, and the Federal Capital Territory (Abuja)—on behalf of the Nigerian people (Federal Republic of Nigeria 2011, p. A 238).

The Act empowers the NSIA to carry out the following functions (Federal Republic of Nigeria 2011):

a. The stabilization of the Nigerian economy during economic stress;
b. Savings for future generations of Nigerians; and
c. Enhancement of infrastructural development in Nigeria.

[4] The membership of the NEC consists of the vice president (as chairman), the 36 State Governors, the governor of the Central Bank of Nigeria, and other co-opted government officials. The National Economic Council (NEC) was established by the provisions of the Constitution of the Federal Republic of Nigeria, 1999, to "advise the President concerning the economic affairs of the Federation, and in particular on measures necessary for the coordination of the economic planning efforts or economic programmes of the various Governments of the Federation."

Unlike the ECA, which served the short-term goal of stabilization, the NSIA was tasked with not only economic stabilization but also the execution of long-term investments to enhance the development of the Nigerian infrastructure and foster intergenerational equity. Nigeria faces a significant infrastructure gap, which has the potential to inhibit its ability to deliver public goods. To achieve these three objectives, the NSIA Act provided for the establishment of three separate ring-fenced investment portfolios: the Stabilization Fund, the Nigeria Infrastructure Fund, and the Future Generations Fund (Federal Republic of Nigeria 2011, p. A228).

In order to proactively respond to its goal of acting as a buffer against fiscal volatility arising from oil price slumps, investments in the Stabilization Funds are targeted at "highly liquid, safe assets" (Orji and Ojekwe-Onyejeli 2017). The Nigeria Infrastructure Fund focuses on long-term investments in domestic infrastructure so as to close Nigeria's yawning infrastructural deficit, which, according to a report by the African Development Bank, requires investment of about $350 billion in ten years to bridge (Oji 2019). To save today's revenue windfall for future generations, the Future Generations Fund was tasked to invest in highly diversified foreign assets (Orji and Ojekwe-Onyejeli 2017). This fund is essential, as DPR estimates put the lifespan of Nigeria's crude oil reserves of 37 billion barrels at 49 years (Asu 2020).

NSIA Funds Allocation

1. Funding of the NSIA

The NSIA Act guarantees consistent allocation of funds to the NSIA to execute its mandates. Section 30 states that subsequent funding of the NSIA shall be derived from Residual Funds above the Budgetary Smoothing Amount of the Federation Account, the primary budget account into which all oil-related revenues are kept and subsequently shared between the three levels of government, "provided that derivation portion of the revenue allocation formula shall not be included as part of this funding" (Federal Republic of Nigeria 2011, p. A237). In other words, the NSIA is to be funded each month from surplus revenues above the annual oil reference price/oil price benchmark in the Federation Account, minus the budgetary smoothing amount and the 13% derivation fund for the oil-producing states.

Following the NSIA's creation, the sum of $1 billion was allocated from the ECA as seed money to the NSIA by the federal, state, and local governments in accordance with a distribution formula for sharing the revenues in the Federation Account included in the separate Allocation of Revenue Act (Orji and Ojekwe-Onyejeli 2017). In 2016, 2017, and 2019, an additional $250 million was provided for the NSIA in each year, bringing the Authority's core capital to $1.750 billion (NSIA 2019b; State House Abuja 2019). In addition to its core assets, the NSIA also manages third-party funds such as the $650 Presidential Infrastructure Development Fund (PIDF), and the $550 million in third-party assets to be managed on behalf of the Nigerian Bulk Electricity Trading PLC (NBET) and the Debt Management Office in the respective sums of $350 million and $200 million (Orji and Ojekwe-Onyejeli 2017; NSIA 2019b). The NSIA's investment assets increased when the United States Department of Justice, the Government of Jersey, and the Federal Republic of Nigeria (FRN) signed an Asset Recovery Agreement to transfer over $308 million of forfeited assets to the NSIA to manage on behalf of Nigeria. The forfeited money was the laundered proceeds of acts of corruption by General Sani Abacha, Nigeria's Head of State from 1993 to 1998, and his family and associates. It will be used to expedite the construction of the three major infrastructure projects across Nigeria: the Lagos to Ibadan expressway, the Abuja to Kano expressway, and the Second Niger Bridge. These projects are currently being executed under the supervision of the NSIA (U.S. Embassy and Consulate in Nigeria 2020).

Third-party funds management aligns with the omnibus provision that gives the NSIA the latitude and flexibility to manage mandates beyond its brief.[5] Section 5(1) of the NSIA Act empowers the Authority to "Undertake, do agree to do activity, incur such expenditures and carry out functions which in the opinion of the Board are necessary, incidental or conducive to the attainment of [its[objects and functions" (Federal Republic of Nigeria 2011, p. A230).

[5] Interview with Uche Orji the NSIA Managing Director/Chief Executive Officer. Interview took place at the NSIA headquarters, Maitama Abuja, on April 24, 2019.

Investment Rules and Strategies

The NSIA Act gives the Authority's Board of Directors the power to allocate capital among the three funds under its management as it deems fit, with the proviso that "each of the three funds may receive no less than 20% of revenues under any allocation" (Orji and Ojekwe-Onyejeli 2017, p. 172). In. line with this proviso, the initial $1 billion in assets allocated to the NSIA was allocated as follows: $400 million each to the Nigeria Infrastructure Fund and the Future Generations Fund and $200 million to the Stabilization Fund (Orji and Ojekwe-Onyejeli 2017).

Table 1 provides a summary of the three funds, their mandates, investment rules, strategies, and performance targets as developed by the NSIA.

The NSIA has added new areas to its original focused investments, such as infrastructure developmentin the midstream sector of the oil and gas industry, including gas pipelines, gas industrialization, oil and gas storage processing and refining facilities, health, and water resources projects. To achieve these expanded goals, it has emphasized the principle of co-investment. For instance, the NSIA is co-investing with Morocco's sovereign fund, Ithmar Capital, to develop a Trans-African gas pipeline that would use Nigeria's vast fossil fuel reserves for energy production for markets in West Africa, North Africa, and possibly continental Europe (Sovereign Wealth Fund Institute 2016, 2019). It is also partnering with Morocco's OCP Group, the world's largest phosphate exporter, to develop Basic Chemicals Investment as part of the Nigerian Presidential Fertilizer Initiative that would lead to the building of a $1.5 billion plant capable of producing 1 million tonnes of ammonia (Chima 2019; Eljechtimi 2019).

Withdrawal Rules

The NSIA is subject to rules regarding how and when the governments can use the mandate funds' income and profits. As the smallest of the three funds under NSIA management designed to be a cash balance, "the Stabilization Fund may be drawn upon if the provisioned budgetary smoothing amount maintained in the Federation Account is insufficient to stabilize the budget and economy owing to oil prices falling below the budgeted price" (Orji and Ojekwe-Onyejeli 2017, p. 173). Given their longer term goals, the law does not envisage regular withdrawal of income

Table 1 NSIA Funds' mandates, investment rules, and targets

	Stabilization Fund	*Infrastructure Fund*	*Future Generations Fund*
% of Assets Under Management	20% allocation	40% allocation	40% allocation
Objective	Provide stabilization support to the Federation revenue in times of economic stress	Invest in a diversified portfolio of growth investments to provide future generations of Nigerians a savings base for such time as the hydrocarbon reserves are exhausted	Enhance the development of infrastructure, primarily through investment in domestic infrastructure projects that meet targeted financial returns
Strategic Asset Allocation/Focus Sectors	Asset class and allocation are: Absolute Return Fixed Income/Investment Grade Corporate Fixed Income (60%); US Treasuries (40%)	Asset class and allocation are: public equity (25%); private equity (25%); absolute returns (25%); and other diversifiers (25%)	Immediate focus sectors are: Real Estate, Healthcare, Power, Agriculture, and Motorways
Mandate/Recent Investments	As at the end of 2013, all capital had been deployed in allocation to three managers: UBS (US Treasury mandate), Goldman Sachs and credit suisse (Corporate Bond mandates)	The asset allocation policy reflects a balance between the Fund's financial and investment objectives, risk tolerance, and need for liquidity	US$ 10 million investment in Fund for Agriculture Finance in Nigeria (FAFIN); US$ 10 million investment in Nigeria Mortgage Refinancing Company (NMRC); Co-developer of the Second Niger Bridge; Project Co-developer of the Nigeria Credit Enhancement Facility with DFID
Expected Avg. Annualized Returns (USD)	1%	5%	6%
Investment Horizon	Short Term	Long Term > 20 years	Long Term > 20 years

Source http://nsia.com.ng/about-us/fund-mandates

from the Future Generations Fund and the Nigeria Infrastructure Fund. Rather, profits, interests, and dividends realized from these funds are to be reinvested in existing or new assets of both funds (Federal Republic of Nigeria 2011).

However, the law empowers the Board to, by unanimous decision, declare a distribution out of "uninvested and uncommitted" available funds of the Authority to the governments of the federation in proportion to their respective contributions (Federal Republic of Nigeria 2011). However, these funds can only be distributed when a net profit has been realized in the fund for at least five years in the year following the Authority's formal establishment (i.e., 2011); if a net profit is recorded by the fund in the year in which such dividend is to be paid; and if the Authority has sufficient funds to meet its own operational needs (Federal Republic of Nigeria 2011, p. A238). The funds are to be paid into the Federation Account, from whence they are distributed to the governments if these conditions are met; finally, the distributed funds should be less than 60% of the profits of the Authority at the time of the distribution (Federal Republic of Nigeria 2011, p. A239).

Governance of the NSIA

Governing Council

The Governing Council is the NSIA's advisory and oversight body, composed of 55 members representing the ownership and stakeholder groups in the fund. This includes the President (who may be represented by the vice president) who chairs the body, 36 state governors, ministers of finance and national planning, the Attorney General of the Federation, the CBN Governor, the Chief Economic Advisor to the President, the Chairman of Revenue Mobilization Allocation and Fiscal Commission (RMAFC), and twelve individuals representing the general public from the private sector, civil society, youth, and academia.

The Council meets at least once a year, receives the performance report of the Authority annually, and, on the basis of this report, raises questions and makes suggestions for improved performance. Most often, the governing council's role is performed by the NEC, which, unlike the council, meets monthly. The NSIA Board presents its account to the NEC, and the NEC approves funding for the Authority or reviews the Authority's performance through an ad hoc committee set up by the governors. However, its prominence as a key decision-making body marginalizes the participation of the nonofficial representatives of the board, such as the members representing civil society, youth, and academia.

Board of Directors

The NSIA's Board of Directors(the Board) is a nine-person body that is vested with the highest operational management responsibility for "the attainment of the objects of the Authority, the making of the policy and general supervision of the management and affairs of the Authority and such other functions conferred upon it by any other provision" of the NSIA Act (Federal Republic of Nigeria 2001, p. A233). The Board consists of three executive and six nonexecutive members, all of whom are appointed by the President on the recommendation of the Minister of Finance (through an independent five-member Ministerial Executive Member Committee constituted by Minister and which includes him/her as a member) in consultation with the NEC.

The Board works through five standing committees: Investment, Risk, Compensation, Audit, and Finance and General Purpose, all of which, in line with the requirement of the Act, are composed of nonexecutive members. While the President appoints members of the Board, this organ of the NSIA is expected to be "independent in the exercise of its responsibilities" (Federal Republic of Nigeria 2011, p. A236). Indeed, Sects. 7(3) and 25(1) of the Act charge the Council to "observe the independence of the Board" and stipulate that "the Council may not, by resolution or otherwise, require the directors to take, or to refrain from taking, any specified action" (FRN 2011, pp. A231, A236).

Although the Board of Directors is the highest organ of the NSIA, the Authority is primarily managed by the Board's three-member team of Executive Directors, or the Executive Management Team. The Executive Management Team has "overall responsibility for the implementation of the establishment of the SWF whose governing principles are based on rules and guidelines that are consistent with international best practices for managing SWFs." The three-member team of Executive Directors is led by the Managing Director/Chief Executive Officer who acts as the Authority's principal officer. According to the NSIA Act, the CEO is responsible for the "execution of the policies formulated by the Board and the day-to-day administration of the Authority" (Federal Republic of

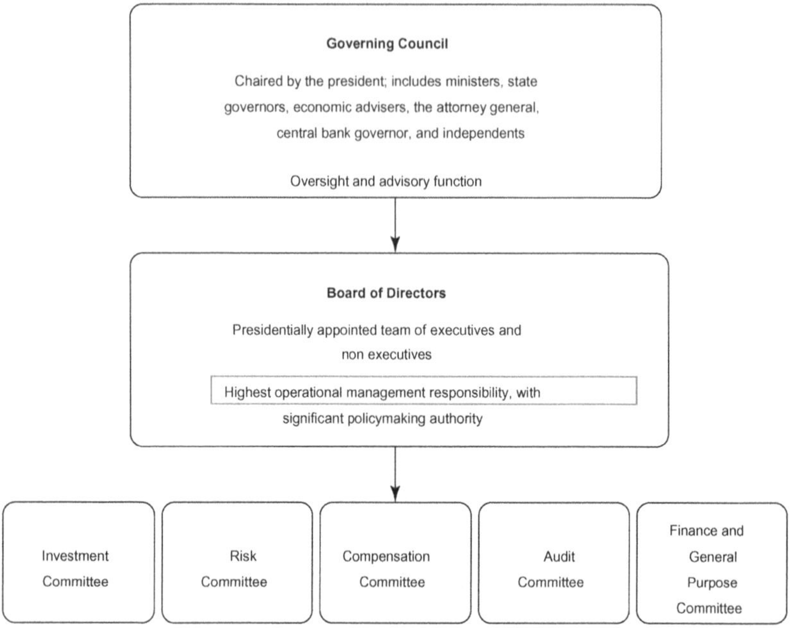

Fig. 1 The governance structure of the Nigeria Sovereign Investment Authority (*Source* Orji and Ojekwe-Onyejeli [2017])

Nigeria 2011, p. A235). The other members of the executive team are the Chief Investment Officer (CIO) and the Chief Risk Officer (CRO) (Fig. 1).

Natural Resource Funds (NRFS) in Operation: Performance, Political Struggles over the Funds, Continuity, and Change

The NSIA has the features of a Natural Resource Fund (NRF) that is designed to be successful. It has clearly defined mandates, and has not yet been enmeshed in scandals of abuse or corruption, or allegations of funds being used for political agendas. Its investments are safe, and the return on investments has been great. For instance, since its creation, and despite the harsh global economic climate, the NSIA has recorded strong growths in several areas such as total assets, investment incomes, profits from investments, and net foreign exchange gains (NSIA 2015, 2016, 2017, 2018, 2019a). It has also adhered to some principles of public accountability as it not only produces audited quarterly and annual reports as stipulated in the NSIA Act, but also makes those annual reports accessible to the public on its website. The NSIA's performance has gained wide acclaim from both the Nigerian government and international/institutional investors. For instance, according to Nigeria's Vice President Osibanjo, "Since inception, the NSIA has maintained the track record of consistent positive performance and displayed an impressive capacity for growth" (Nwabughiogu 2017). In 2018, Nigeria was ranked number 1 in the first edition of the African Sovereign Wealth Fund Index, which ranked Africa's 12 SWFs (Nigerian Investment Promotion Commission 2018). Also, the NSIA has consistently ranked close to the highest (9/10) and was the best ranked in Africa in the Linaburg-Maduell Transparency Index, developed by the Sovereign Wealth Fund Institute (SWFI) to measure compliance with international best transparency practices in such areas as the provision of up-to-date independently audited annual reports, clear strategies and objectives, and management of their own website (Sovereign Wealth Fund Institute [n.d.]). The NSIA's governing structure is central to its performance. This structure is designed such that it is immunized to political interference. As the Wilson Centre (2016, p. 3) noted, the NSIA "is not part of the government's fiscal framework, but rather is managed by an independent and highly engaged oversight board with demonstrable professional capital investment experience. This structure has proven critical for maintaining institutional integrity and enforcing clear rules about when and how the government can request funds from the SWF."

Despite its impressive record, the NSIA was bedeviled from the beginning by the same opposition that the ECA faced as a fund owned by governments of the Nigerian federation. These issues have undermined not only the forging of the political consensus required to fund the NSIA as required by the law, but also the dismantling of the ECA and its takeover by the NSIA. As Adio Waziri of the NEITI noted, "Our paltry oil savings defeat the rationale for having such savings in the first place. Nigeria does not have enough oil savings to finance even a fifth of a year's budget at the federal level, not to talk of having enough for investments or for the future generation" (NEITI 2017, p. 14). Rather than having one fund as was envisaged with the creation of the NSIA or the expectation that the ECA would be abolished and taken over by the NSIA (*African Review* 2012, p. 2), Nigeria has two funds existing simultaneously, with the ECA, which was expected to have been folded up and its funds taken over by the NSIA, serving as the primary account. Accordingly, since its creation in 2011, the NSIA has received only $1.75 billion as its core capital. Accordingly, even the modest objective, expressed at its founding by then-Finance Minister Ngozi Okonjo-Iweala, to grow the fund to about $5bn to $6bn in the future (Blas, 2013) has not taken place after eight years.

The political opposition has not only delayed the take-up of the NISA—the Authority's Board was inaugurated 15 months after the NSIA Act was given presidential assent—but has also prevented the consistent allocation of funds to the NSIA as stipulated in the Act. Accordingly, Nigeria's "spend-it-all or even save – and spend" oil wealth mismanagement culture (NEITI 2017, p. 15) has not been helped much by the creation of the NSIA. As the Minister of Finance noted, while "the NSIA is doing a good job with the small savings available to it in the Stabilization Account," the government has not been able to attain the "provision in the NSIA Act that requires that savings in the ECA on a routine basis be able to fund the NSIA" (Vanguard 2019).

The following section discusses the issues that affect the operation of the NSIA in their historical context, highlighting the mechanisms and processes through which they have undermined or can weaken the robust and sustainable realization of the Authority's objectives.

The constitutional question is central to the political contestations surrounding the NSIA. Not long after the NSIA was created, the ostensible compromise between the federal and state governments over its creation began to unravel. A major deficiency of the NSIA, which has provided the foundation for political opposition, was its shaky constitutional foundation. Like that of the ECA, the establishment of the NSIA was fraught with contestations from state governors who saw the provision mandating transfer of surplus oil revenues to the NSIA as a diversion of revenues meant for the states, and a violation of Sect. 162 of the 1999 constitutional provision mandating that all revenues must be shared between the governments of the federation. To reiterate, Sect. 162(1) of the constitution provided:

> The Federation shall maintain a special account to be called "the Federation Account" into which shall be paid all revenues collected by the Government of the Federation, except the proceeds from the personal income tax of the personnel of the armed forces of the Federation, the Nigeria Police Force, the Ministry or department of government charged with responsibility for Foreign Affairs and the residents of the Federal Capital Territory, Abuja.

However, the opposition of the governors began even before formal legislative efforts aimed at establishing the NSIA. In June 2010, following President Jonathan's declaration in April of that year that he was interested in establishing a sovereign wealth fund that would replace the ECA, the state governors who had benefited from the ECA during the global economic crisis maintained that they wanted to retain the ECA, which then stood at around $3.4 billion, and opposed the use of the ECA to fund the proposed sovereign wealth fund (Onuah 2010a). According to Suswan, who spoke for his colleagues, "We the governors think there should be an alternative source of funding to be used (for the sovereign wealth fund) than that of the excess crude account." This opposition seemed to subside when the NEC, consisting of governors and the President, approved the creation of a sovereign wealth fund in the same year. Yet, this seemingly cooperative attitude soon dissipated when the governors renewed their opposition to the NSIA after the bill that established it was passed by the National Assembly and signed by President Jonathan in May 2011. In August 2011, before the NSIA even took off, the Nigerian governors' forum "called on the federal government to suspend implementation of the Act pending resolution of some undisclosed issues." (Ukiwo et al. 2012, p. 163). Thus, even though they were part of the NEC's decision to create the NSIA, the governors rescinded that decision and withdrew their support immediately after the Authority was created as an Act of the National Assembly. In October 2011, the governors made their opposition legal by approaching the Supreme Court to prevent the withdrawal of $1 billion from the ECA as seed money for the NSIA, citing the same reason they had adduced for their opposition to the ECA: that any transfer of money to an account other than the Federation Account is illegal and unconstitutional.

Although consensus had been generated for the SWF through the NEC, which has the 36 state governors as members, the governors have not been able to sustain this consensus as they disavowed their support almost immediately. For instance, in June 2012, the governors reached a compromise with the federal government, agreeing that the sovereign wealth fund should start with just $1 billion instead of $7 billion in the ECA (Eboh 2012a, b). The governors themselves confirmed that with this compromise, they had agreed to support the NSIA, reversing their previous decision to block the creation of the fund. According to Rotimi Amaechi, the Chairman of the Nigerian Governors Forum (NGF), "The National Economic Council has agreed with the Federal Government to go ahead to implement the Nigeria Sovereign Investment Authority Act with an initial fund of $1 billion" (Ogidan and Shadare 2012). However, it was not long before the governors jettisoned this cooperation and went back to the court.

The governors seemed to be on a firm footing as the 2002 Supreme Court's judgment *Attorney General, Ogun & Others vs. Attorney General, Federation* had declared the federal government's practice of non-remittal of revenues from natural gas into the Federation Account illegal. This reality forced the federal government to ask for an out-of-court settlement when the case came up at the Supreme Court on several occasions, despite the federal government's insistence that it had a legal basis to create the SWF.[6] This court case is still lingering, even though it has not been activated in recent years.

As the legal case became protracted, the governors used their superior legal standing and political pressures to ensure the continuing existence of the ECA instead of the original plan devised by then-Finance Minister Ngozi Okonjo-Iweala to replace the ECA with a longer-term investment vehicle such as the NSIA (Eboh 2012a). Thus, in a paradoxical move, the governors who had been campaigning for the abolition of the ECA because of its illegality decided to support its continuation rather than replacing it with the newly created NSIA. The reason for this is that, unlike the NSIA, which is protected from indiscriminate drawdowns, the ECA is a fund with no legal backing and codified withdrawal and spending rules. The ECA is also flexible, allowing the governors to withdraw funds

[6] The federal government argued that the consensus of the legal experts it consulted was that the creation of a SWF would not contravene Sect. 162 of the constitution mandating distribution of money from the Federation Account among the three levels of government, as the money put into the SWF is essentially a distribution, albeit a deferred one, and that share certificates would be issued to the states as evidence of their investment in the SWF (Ajayi and Olaleye 2011).

from it through informal negotiations with the President as they have done in the past, especially when the President is politically vulnerable, for instance, during elections when the presidents need the governors' support.

Therefore, even though the NSIA Act envisaged that surplus oil revenues should be transferred every month from the Federation Account to the NSIA, the governors capitalized on their leverage to ensure these revenues are still transferred to the ECA from where they can decide how much, and when, money goes to the NSIA, thereby starving the NSIA of funds. For instance, since the agreement for the transfer of only $1 billion seed money to the NSIA in 2012, no additional core allocation was made until 2016 (NSIA 2019b).

The National Assembly has also been implicated in the discord over the ECA and NSIA, though not as much as the governors and presidents. It was sidelined by President Obasanjo when the ECA was created, but played a central role in approving the establishment of the NSIA in 2011. Following the creation of the NSIA, the NA became preoccupied with the fate of the two funds. In 2017, in the face of repeated withdrawals from the ECA and reports by the NRGI that the ECA was the world's second worst-governed fund (Akpan and Ejoh 2017), the Senate called on the federal government to abolish it, deposit the surplus oil revenues into the Federation Account as stipulated by the constitution, and transfer some to the NSIA (Akpan and Ejoh 2017; Umoru 2017). In introducing the bill for the abolition, Sen Rose Oko noted, "These breaches of the constitution in setting up and operating the ECA have created room for the pool of funds from revenue accruing to the federation being operated without legal backing" (Akpan and Ejoh 2017), and that this situation has made the fund into the "biggest slush fund" for the executive, particularly the governors (Akpan and Ejoh 2017; Umoru 2017). However, the motion for the probing of the use of ECA funds was defeated, as the 8th Senate where this motion was tabled had sixteen members who were governors and eight who were deputy governors during the period of the proposed probe; these members lobbied other senators to reject the probe, which would have also implicated them (Umoru 2017).

In spite of the Senate resolution, however, the ECA was not scrapped. The intervention of the legislators continued with the resolution of the House of Representatives, in February 2018, to propose its own solution to the ECA challenge. The House of Representatives attempted to bring the ECA under its control by legal backing, and hence legalize the

ECA to protect it from persistent raiding by executives of the federal and state governments. This resolution of the House of Representatives, mandating its finance committee to ascertain the status of the ECA, came in view of the National Economic Council's decision to contribute $1 billion for the fight against Boko Haram in January 2018. Such use of money from the ECA for counter-terrorism was criticized in some quarters, including some governors who had approved the decision as members of the NEC. The attempt to bring the ECA under the legislative control of the NA was therefore meant to give it formal status and prevent indiscriminate withdrawals. The proposal for the legalization of the ECA involved an amendment of the extant Allocation of Revenue (Federation Account etc.) Act, Cap. A15, Laws of the Federation of Nigeria, 2004, as follows: "There is hereby established for the federation an Excess Revenue Fund Account, which shall consist of all revenues or other money raised or received by the federation above the revenue targets set out for the purpose of funding the budget in a fiscal year" (Ameh 2018). However, the attempt to legalize the ECA ran into the same problem of constitutionality that the ECA itself has had to confront. The key issue that undermined this proposal was the question of "whether another account outside the Federation Account recognised by the constitution could be created by a bill as against amending the constitution" (Ameh 2018). Based on this doubt, the proposal to promulgate a new law transferring legislative power over the ECA to the National Assembly was not considered a bill, and the ECA continued to exist simultaneously with the NSIA, rather than being absorbed by the latter or abolished completely.

While the issue of the constitutionality of the ECA, and later the NSIA, was the central sticking point in the opposition of governors to the funds, the constitutionality question was a rallying cry for other broader political economic issues. First is the issue of trust in federal-state relations. According to Rothstein (2005), it is difficult for people to efficiently cooperate for common purposes if they doubt that most other people will also choose to cooperate. One such example of an issue in which the lack of trust in the federal government by states has played a pernicious role is the states' quest for the complete sharing of the Excess Crude Account (ECA), even when they knew that doing so would lead to the depletion of the fund. The states are afraid that with the federal government's attitude of unilaterally spending money from the ECA, the government could drain the account to finance its budget deficits, especially with the sharp decline in the price of oil, to the detriment of the states. The collective memory that stimulates this thinking is the fact that the ECA, critical as it is, circumvented due process and constitutionalism as it was not established by then-President Obasanjo through the National Assembly. The unilateral withdrawals of funds from the account, which Gillies has described as "ad hoc and discretionary" (2010, p. 2), are also worrisome. President Obasanjo withdrew funds from the ECA for the federal government's own projects such as the payment to Paris Club creditors to eliminate Nigeria's debt, payments for contracts under the National Integrated Power Projects (NIPPs) and the Railway Modernization and Expansion Project, and for additional funding for a two-day extension of the 2006 Census. Obasanjo has also been accused of using funds from the ECA to finance the aborted bid to extend his tenure beyond the constitutional two-term limit (1999–2007), and some federal legislators were allegedly offered the part of the money as bribes to support the Bill (Odinkalu and Osori 2018).

Successive presidents also made similar discretionary withdrawals, even though some of these were used to appease the governors. Then-Governor Tunde Fashola put the issue of trust in perspective when he noted that:

> It is not that the governors are up in arms against the idea of saving. But we are asking what the rules of engagement are and do those rules of engagement work within the rules that bind all of us? Before you save on my behalf, there is also need to address the issue of trust. How efficiently have you managed the funds that the federation has put in your trust? And what makes you the better saver and better investors? And is the saving done within an expectable framework of the constitution? Those are the issues surrounding the Sovereign Wealth Fund (Eboh & Ibeabuchi 2012).

The politics of federal-state relations, especially the federal government's lack of credible commitment to its own fiscal reforms (Okpanachi 2010), as demonstrated by its unilateral withdrawals from the ECA, created a high level of mistrust about the real intentions of the federal government's reform aimed at saving excess oil revenues in a new sovereign wealth fund.

A second concern is electoral turnover, particularly the disincentive to sustainable savings by incumbent politicians who are fixated on immediate gains that would benefit them rather than their successors. Political scientist Paul Pierson has underscored the incentives provided by electoral turnover for politicians to discount the long-term consequences of decision-making in favor of short-term goals. According to Pierson, even though outcomes of political actions and complex institutional reforms "only play out in the long run ... the long run, "political actors, especially politicians are often most interested in the short-term consequences of their actions" while the "long-term effects are heavily discounted" (2012, p. 41). This insight can also help us understand the opposition of the governors to the creation of the NSIA, even though they recommended its creation through the NEC. The NSIA Act was signed in May 2011, just a month after the general election of that year. With the NSIA stipulating that withdrawals from the NSIA funds would be allowed only after five years of profitable investments, governors recently elected to their first four-year term, or those who were returned for a second and last term, felt that the benefits of the NSIA may accrue not to them but their successors if or when their tenure ended in the electoral cycle of 2015 (Oxford Business Group 2012).

The third issue is that of control over the NSIA. A key factor that fueled the governors' opposition to the NSIA and their decision to approach the Supreme Court to stop its establishment was their allegation that, since they gave approval to establish a SWF in 2010, they were "not fully carried along during the conceptualization stage and subsequent signing of the NSIA Act" (Ogidan and Shadare 2012). Following the appointment of Ngozi Okonjo-Iweala as the new Finance Minister in June 2011, a compromise was reached with the governors to address their concern of marginalization in the decision-making process of the proposed SWF. The governors secured a concession that made all 36 state governors members of the NSIA Governing Council "instead of the six representatives drawn from the six geopolitical zones that was initially recommended to seat on the council" (Ajayi and Olaleye 2011).

According to the NSIA Act, "While state governors are to be represented on the NSIA's governing board, ultimate authority would rest with the federal government" (Oxford Business Group 2012, p. 51). To be sure, the governing council has a purely advisory role. However, it can approve transfers to the NSIA, and the governors have used their membership of this body to shape decisions regarding allocation of funds to the NSIA. However, even though the governors found a way to navigate the issue of composition of the Governing Council, they were still not happy that they do not have effective influence over the executive board, which is "the only body with the power to veto NSIA decisions" (Oxford Business Group 2012). They were also unable to influence the choice of the executive team, as audit firm KPMG was contracted to carry out the recruitment exercise.

Finally, at the core of the problems undermining the NSIA is the political economy of oil in Nigeria's fiscal federalism; in other words, the overdependence of governments on oil revenues. With this reality in mind, even if a consensus for regular transfer of money to the NSIA is achieved, such a regular transfer would be difficult in practice. This is because, as it is presently configured, the NSIA would not be funded when oil prices are low, even if the law for regular funding is adhered to. Because the ECA relies overwhelmingly on oil windfalls, the account will remain unfunded when oil prices plunge below the oil price benchmark in the yearly Appropriation Act. The Nigeria Extractive Industries Transparency Initiative (NEITI) has advised the "government to ensure constant savings whether oil prices are high or low" (NEITI 2017). However, as attractive as this recommendation is, its practical workability remains in doubt given the governments' reliance on oil. With states that overwhelmingly depend on oil, attempts to save their share of oil windfalls in a permanent vehicle are viewed as attempts to confiscate or hoard money meant for development.

Although non-oil revenue has been increasing in recent years, and currently accounts for 30% of total revenue (Francis 2015), Nigeria's fiscal structure is still heavily slanted in favor of dependence on oil revenues. This means that intergovernmental distribution of oil revenues from the Federation Account is still central to the survival of these states. The fact that most of the states cannot fund their budgets without federally shared revenues[7] means that the state governors' fight over these revenues was not only about the constitution, but also the economic viability of the states, and hence the political survival of the governors.

Part of the reason that states overwhelmingly depend on transfers from the Federation Account is that they have blatantly ignored the development of their own source or internal revenues. This is large because of the political economy of oil, which encouraged the creation of states with guaranteed revenues from the Federation Account on political grounds without considering their economic viability.

[7] According to a recent document, apart from Lagos, Rivers, and Akwa Ibom states, Nigeria's other "thirty-three state governments cannot finance their recurrent expenditure without allocation from the federation account" (Onuba 2019).

The reality of Nigerian states transferring dependence (Asadurian et al. 2006), especially in the face of increasing citizens expectation and challenges, led the governors to oppose saving their shares of oil windfalls in a SWF, which they believe hinders much-needed revenues by tying them up for development. According to one governor: "We cannot say we are saving money when people are hungry. You cannot save money when your child is in the hospital. If your child dies you are a fool (Ezeamalu 2015)."

Another similarly noted:

> The thing is that it is our money and we need it. We are faced with so many challenges beyond the scope of our financial capacity and we cannot have a situation where the federal government will compel us to save money by not giving us our full dues from the Federation Account. For example, we need money to develop our region, how do we get that? We are confronted daily with security and conflict resolution matters, yet we don't have enough resources to tackle them. (Ajayi and Olaleye 2011)

The ballooning of already-bloated government wages has also put pressure on the governors. In February 2011, the Nigerian Senate approved an increase in the minimum wage from about $50 to $120, with some governors initially claiming that their state revenues would not be able to pay and sustain the new salary increment. In fact, the governors had insisted, following the meeting of the NGF in June 2010, that "paying the extra wage would make their states broke and paralyzed" (Sahara Reporters 2011). They argued that they would only be able to pay if the existing revenue-sharing formula were reviewed to increase their monthly statutory allocations from the Federation Account. The governors were forced to agree to pay the new wage following threats of industrial action by the labor unions. Indeed, the sudden volte-face of the governors who had earlier approved the creation of a SWF (as members of the NEC) can be partly explained by the financial burden of implementing the National Minimum Wage (Ukiwo et al. 2012, p. 163).

Yet, it would be naïve to think that the governors oppose the transfer of excess oil money to the SWF only because they need money to invest in development and pro-people projects. Reduction of the money that goes to the state does not only mean a weakening of their ability to deliver on electoral promises, but more importantly means a reduction in the luxurious lifestyles they enjoy (at the expense of the citizens), the huge patronage they wield, and oil rents to sustain their grips on power. In a federal system in which state and local government revenues are automatically guaranteed from the Federation Account, it is not surprising that the use of revenues from oil extraction by the subnational governments continues to tilt toward opportunistic goals such as corruption, recurrent expenditure such as payment for political appointees, or consumption, rather than the provision of public goods (Okpanachi 2019).

A final issue with the NSIA is that even though its Establishment Act stated that the governments of the federation are mere custodians expected to hold the funds in trust for the Nigerian people as owners of the resources, the "people" aspect has been downgraded in its governance. The Act does recognize the importance of the people by mandating the appointment of individuals representing the public, including youth, academia, and civil society representatives as members of the NSIA's governing board. However, this has not taken place,[8] and so members representing governments, especially state governors and the president (usually represented by the vice president), have been the key members of the governing council. Therefore, it is taken for granted in the current governance of the NSIA that the Authority manages government-owned funds, even though one of the funds is focused on saving for future generations.

Although it could be argued that the publication of the Authority's reports, which are made available to the public, ensures accountability to the public, the lack of public input into the Authority's decision-making is a yawning lacuna. This gap is heightened by the fact that civil society itself has not shown serious concern with the agency.

Conclusion

The NSIA started with a challenge, as the state governments initially opposed its establishment and funding from surplus revenues above the oil benchmark. However, the initial opposition soon dissipated following a compromise between the federal and state governments to maintain the ECA in parallel with the NSIA, rather than the latter replacing the former as originally planned. Yet, though the NSIA has avoided the constant

[8] Interview with Uche Orji the NSIA Managing Director/Chief Executive Officer. ibid.

political conflict and pressures for sharing of money that has characterized the ECA, the elite consensus that retained the ECA meant that funds intended for the NSIA would go to the ECA instead. With surplus oil funds domiciled in the ECA, where they are vulnerable to ad hoc and discretionary withdrawals, rather than the NSIA which is designed to save funds for the future, Nigeria has continued to lack both a buffer against oil prices and a stable fund for future generations. As of March 2020, the ECA held just $71.8 million (Iroanusi 2020).

From the evidence so far, the NSIA can be described as a well-designed and technically managed fund with little or no political interference, albeit not without some limitations. One major limitation highlighted in this chapter is the absence of civil society participation in the governance of the Authority. This is an important deficit, especially considering that its Establishment Act explicitly stated that the ultimate owners of the fund are the Nigerian people. The lack of civil society participation from within, and limited public debate and scrutiny or criticism from outside, meant that the Authority has been operating without a critical check and voice that could inform the refinement of ideas and decisions or pressure the government to build up savings in the sovereign wealth fund. This gap notwithstanding, the general verdict is that the NSIA design and operation is a model for other countries. The real problem, therefore, is the inability to fold the ECA into the NSIA as originally envisaged. The ECA continues to serve as the primary holder of Nigeria's surplus oil revenues, creating uncertainty about the transfers to the NSIA. With the dramatic slump in oil prices from a high of $62 per barrel in November 2019 to $20 per barrel in March 2020 as a result of dampened fuel demand arising from the coronavirus pandemic and supply glut due to the oil price war between Russia and Saudi Arabia (Egan 2020), the almost-depleted ECA itself will not be funded, with the implication that the NSIA's low assets will not increase in the foreseeable future.

Nigeria weathered the 2008–2009 global crisis using savings in the ECA, which stood at $20 billion in 2009 (Akinkugbe-Filani 2020), without resorting to IMF/World Bank loans. By contrast, in 2020, Nigeria's informal oil fund (the ECA) and the formal one (NSIA) do not have sufficent funds to enable the country to weather the storm. With a budget that is based on an oil price of $57 per barrel, the economy, which has yet to completely overcome the strain of the 2014 oil price collapse, will be in even more dire financial stress in 2020, and investments in the NSIA, even without the ECA challenge, will remain a mirage. Ironically, with declining oil revenues which has forced the government to review the budget's oil benchmark from the original $57 per barrel to $30 per barrel, plan has been finalized for the withdrawal of $150 million from the NSIA for distribution to the three tiers of government to help tackle the impact of the coronavirus pandemic on the economy (Channel TV 2020), even though little was saved in the fund when oil prices were high. The inability to save for the long term in the ECA or ensure the replacement of the ECA with the NSIA is intricately framed around questions over the constitutionality of such savings, and the politics of entitlement woven around the constitutionality question. However, its deeper mechanism/driver is Nigeria's overdependence on oil. Even though non-oil revenues have increased in recent times, "the naira is still a classic petrocurrency whose fate remains intrinsically tied to global oil prices" (Akinkugbe-Filani 2020). Accordingly, even though the NSIA has been innovative in utilizing its current allocation, the uncertainty of revenue inflows to the NSIA or the absence of guaranteed funding for the NSIA will continue, thereby undermining the achievement of its goals and the realization of its mandates.

References

Abdallah N, Adugbo D (2016) Nigeria loses N1.1tr to militant attacks. Daily Trust (Abuja), 29 Aug. www.dailytrust.com.ng/news/general/nigeria-loses-n1-1tr-to-militantattacks/160529.html

Adeniyi O (2015) Buhari and the Sovereign Wealth dilemma. ThisDay, 22 Oct. http://www.newsexpressngr.com/news/16581-Buhari-and-the-Sovereign-Wealth-Dilemma-By-Olusegun-Adeniyi

Adugbo D (2019) Mixed feelings over oil discovery in the north. Daily Trust, 19 Oct. https://www.dailytrust.com.ng/mixed-feelings-over-oil-discovery-in-the-north.html

African Review (2012) Iweala-Okonjo's transformation agenda. 28 Sept. www.africanreview.com/finance/economy/iweala-okonjo-s-transformation-agenda?start=1

Akpan E, Ejoh E (2017) $2.4bn ECA one of world's most poorly governed SWFs —NRGI Read more at: https://www.vanguardngr.com/2017/06/2-4bn-eca-one-worlds-poorly-governed-swfs-nrgi/

Akinkugbe-Filani R (2020) How coronavirus and the global oil price war can impact Nigeria. The Africa Report, 11 Mar. https://www.theafricareport.com/24442/how-coronavirus-and-the-global-oil-price-war-can-impact-nigeria/

Alli Y (2012) Peace moves on N2.29tr Excess Crude Account. The Nation, 29 Aug. https://www.proshareng.com/news/Nigeria%20Economy/Peace-moves-on-N2.29tr-Excess-Crude-Account/18100
Ameh J (2018) Reps move to legalise Excess Crude Account. Punch. 21 Feb. https://punchng.com/reps-move-to-legalise-excess-crude-account/
Ajayi Y, Olaleye O (2011) Governors Seek To Stop Sovereign Wealth Fund. This Day, 21 Aug. https://www.nairaland.com/741104/governors-seek-stop-sovereign-wealth
Asadurian T, Nnadozie E, Wantchekon L (2006) Transfer dependence and regional disparities in Nigerian federalism. In: Wallack J, Srinivasan T (eds) Federalism and economic reform international perspectives. Cambridge UP, Cambridge, p 407–455
Asu F (2020) Nigeria's oil reserves'll run dry in 49 years—DPR. Punch. 20 Feb. https://punchng.com/nigerias-oil-reservesll-run-dry-in-49-years-dpr/
Auty R (1993) Sustaining development in mineral economies: The resource curse thesis. Routledge, London
BBC News (2018) Boko Haram releases 13 hostages to Nigerian government. 10 Feb. https://www.bbc.com/news/world-africa-43018285
Blas J (2013) Protecting Nigeria oil SWF is no easy task. Financial Times, 10 Oct. https://www.ft.com/content/69325c28-2440-11e3-a8f7-00144feab7de
Central Bank of Nigeria (2012) The Sovereign wealth fund: Understanding monetary policy Series No. 18. Central Bank of Nigeria, Abuja
Chima O (2019) Nigeria: At $1.5bn, Nigeria lags behind peers in SWF savings. Thisday. 17 June. https://allafrica.com/stories/201906170285.html
Eboh C (2012a) Nigeria picks team for $1-billion fund. The Globe and Mail, 28 Aug. https://www.theglobeandmail.com/report-on-business/international-business/african-and-mideast-business/nigeria-picks-team-for-1-billion-fund/article4505765/
Eboh C (2012b) Nigerian state governors to block oil wealth fund. The Globe and Mail, 20 Sept. https://www.theglobeandmail.com/report-on-business/international-business/african-and-mideast-business/nigerian-state-governors-to-block-oil-wealth-fund/article4557965/
Eboh M, Ibeabuchi C (2012) Why govs oppose Sovereign Wealth Fund – Fashola. Vanguard, 17 May. https://www.vanguardngr.com/2012/05/why-govs-oppose-sovereign-wealth-fund-fashola/
Egan M (2020) Oil collapses by another 24% to $20. It hasn't been this low since 2002. 18 Mar. https://www.cnn.com/2020/03/18/business/crude-oil-prices-coronavirus/index.html
Eljechtimi A (2019) Update 1-Morocco's OCP plans African chemical plants, fertilizer blenders. Reuters, 4 Mar. https://www.reuters.com/article/morocco-phosphate/update-1-moroccos-ocp-plans-african-chemical-plants-fertiliser-blenders-idUSL5N20R4TG

Ezeamalu B (2015) Governors asked Jonathan to share Nigeria's foreign reserves—Akpabio. Premium Times, 9 Jan. https://www.premiumtimesng.com/news/top-news/174554-governors-asked-jonathan-share-nigerias-foreign-reserves-akpabio.html

Federal Republic of Nigeria (1999) 1999 Constitution. Federal Ministry of Information, Abuja

Federal Republic of Nigeria (2011) Nigeria Sovereign Investment (Establishment etc.) Act, 2011. Official Gazette No. 64, Vol. 98. 3 June. https://nsia.com.ng/~nsia/sites/default/files/downloads/NSIA%20Act.pdf

Francis N (2015) Nigeria: Okonjo-Iweala—Incoming administration must build consensus on fiscal buffers. ThisDay, 16 Apr. https://allafrica.com/stories/201504160329.html

Gillies A (2010) Sovereign wealth fund requires legal standing, binding rules and transparency. https://resourcegovernance.org/sites/default/files/SWF%20recommendations_FINAL.pdf

Iledare W, Suberu R (2012) Nigeria. In: Anderson G (ed) Oil and gas in federal systemsOxford UP, Don Mills, p 227–256

International Monetary Fund (2019) Nigeria: IMF Country Report no. 19/92. https://www.imf.org/en/Publications/CR/Issues/2019/04/01/Nigeria-2019-Article-IV-Consultation-Press-Release-Staff-Report-and-Statement-by-the-46726

Iroanusi Q (2020) ECA: Nigerian govt pays lawyer $4 million "consultancy fee"; Senate wants identity of recipient. Premium Times, 10 Mar. https://www.premiumtimesng.com/news/headlines/381205-eca-nigerian-govt-pays-lawyer-4-million-consultancy-fee-senate-wants-identity-of-recipient.html

Kay C, Wallace P (2016) Oil-rich Nigeria now can't afford to pay teachers. Bloomberg, 18 Feb. https://www.bloomberg.com/news/articles/2016-02-17/nigeria-faces-double-whammy-as-buhari-fails-to-manage-oil-rout

Lewis P (2007) Growing apart: Oil, politics and economic change in Indonesia and Nigeria. U of Michigan P, Ann Arbor.

Madujibeya S (1976) Oil and Nigeria's Economic Development. African Affairs 75(300):284–316

Murray S, Akpe S (2006) FG proposes N136bn supplementary budget. https://onlinenigeria.com/nm/templates/?a=8605

Nigerian National Petroleum Corporation (2010) History of the Nigerian petroleum industry. www.nnpcgroup.com/NNPCBusiness/Businessinformation/OilGasinNigeria/IndustryHistory.aspx

Nigerian Investment Promotion Commission (2018) Nigeria Tops African Sovereign Wealth Fund Ranking. 7 June. https://www.nipc.gov.ng/2018/06/07/nigeria-tops-african-sovereign-wealth-fund-ranking/

NEITI (2017) Transfer oil revenue savings to NSIA–NEITI. https://www.neiti.gov.ng/phocadownload/July%20Newsletter%202017.pdf

NSIA (2015) Annual Reports & Accounts 2014. https://nsia.com.ng/~nsia/sites/default/files/annual-reports/NSIA%202014%20Annual%20Report.pdf

NSIA (2016) NSIA Annual Report 2015. www.nsia.com.ng/~nsia/sites/default/files/annual-reports/NSIA%202015%20Annual%20Report.pdf

NSIA (2017) Annual Report & Accounts 2016. www.nsia.com.ng/~nsia/sites/default/files/annual-reports/NSIA%202016%20Annual%20Report.pdf

NSIA (2018) Annual Reports & Accounts 2017. https://nsia.com.ng/~nsia/sites/default/files/annual-reports/NSIA%20Annual%20Report-2017%20%281%29.pdf

NSIA (2019a) Annual Reports & Accounts 2018. https://nsia.com.ng/~nsia/sites/default/files/annual-reports/NSIA%20Annual%20Report%202018.pdf

NSIA (2019b) NSIA Announces Audited Financial Results for 2018 Financial Year. https://www.nsia.com.ng/~nsia/sites/default/files/press-release/Press%20Release%20-%202018%20Financial%20Statement_0.pdf

Nwabughiogu L (2017) Osinbajo inaugurates Nigeria Sovereign Investment Authority, NSIA, board. Vanguard, 12 May. https://www.vanguardngr.com/2017/05/osinbajo-inaugurates-nigeria-sovereign-investment-authority-nsia-board/

Obasanjo O (2005) Why we must be alert and vigilant, by Obasanjo. 27 July. https://onlinenigeria.com/nm/templates/?a=4076

Obiejesi K (2010) Auditor-General says Excess Crude Account is illegal. https://www.icirnigeria.org/auditor-general-says-excess-crude-account-is-illegal/

Obiejesi K (2018) Auditor-General says Excess Crude Account is illegal. International Centre for Investigative Reporting, 20 July. https://www.icirnigeria.org/auditor-general-says-excess-crude-account-is-illegal/

Odinkalu C, Osori A (2018) Too good to die. Kachifo Limited, Yaba, Lagos

Ogidan A, Shadare W (2012) Analysis: Nigeria's Sovereign Wealth Fund. The Guardian, 28 June. https://www.proshareng.com/news/NIGERIA%20ECONOMY/Analysis--Nigeria's-Sovereign-Wealth-Fund/17588

Oji H (2019) How to tackle Nigeria's huge infrastructure deficit. The Guardian, 10 Dec. https://guardian.ng/business-services/how-to-tackle-nigerias-huge-infrastructure-deficit/

Okpanachi E (2010) Federalism and economic growth: The importance of context in Nigerian public finance reform. Publius: The Journal of Federalism 41(2):311–335

Okpanachi E (2019) The limits of federation-wide political parties in ensuring federal stability: Nigeria under the Peoples Democratic Party. Publius: The Journal of Federalism 49(4):617–641

Onoh J (1983) The Nigerian oil economy: From prosperity to glut. St. Martins Press, New York

Onuah F (2010a) Nigerian governors oppose sovereign wealth fund. Reuters, 15 June. https://uk.reuters.com/article/nigeria-fund/nigerian-governors-oppose-sovereign-wealth-fund-idUKLDE65E2GS20100615

Onuah F (2010b) UPDATE 2-Nigeria president submits bill to create wealth fund. Reuters, 13 Sept. https://www.reuters.com/article/nigeria-fund-idINLDE68C1E220100913

Onuba I (2019) 33 states can't survive without federal allocation—Report. Punch, 24 Oct. https://punchng.com/33-states-cant-survive-without-federal-allocation-report/

Onuoha F (2016) The resurgence of militancy in Nigeria's oil-rich Niger Delta and the dangers of militarization. Aljazeera Centre for Studies, 8 June. studies.aljazeera.net/en/reports/2016/06/resurgence-militancy-nigerias-oil-rich-niger-delta-dangersmilitarisation-160608065729726.html#a4

OPEC (2020) Nigeria facts and figures. https://www.opec.org/opec_web/en/about_us/167.htm

Orji U, Ojekwe-Onyejeli S (2017) The role of the Nigeria Sovereign Investment Authority in a new era of fiscal responsibility. In: Rietveld M, Toledano P (eds) The new frontiers of sovereign investment. Columbia UP, New York, p 168–182

Oxford Business Group (2012) The Report, Nigeria 2012. Macro-level management: A sovereign wealth fund is in the process of being set up. https://oxfordbusinessgroup.com/analysis/macro-level-management-sovereign-wealth-fund-process-being-set

Pierson P (2004) Politics in time: History, institutions and social analysis. Princeton UP, Princeton

Pinto B (1987) Nigeria during and after the oil boom: A policy comparison with Indonesia. The World Bank Economic Review 1(3):419–445

Rothstein B (2005) Social traps and the problem of trust. Cambridge UP, Cambridge

Sahara Reporters (2011) Governors bow to labour, agree to pay 18000 Naira new minimum wage. 16 July. saharareporters.com/2011/07/16/governors-bow-labour-agree-pay-18000-naira-new-minimum-wage

Shehu M (2010) Nigeria: Jonathan Submits Bill to Create Wealth Fund. Daily Trust, 14 Sept. https://allafrica.com/stories/201009140505.html

Sovereign Wealth Fund Institute (n.d.) Linaburg-Maduell Transparency Index (LMTI) https://www.swfinstitute.org/research/linaburg-maduell transparency-index

Sovereign Wealth Fund Institute (2016) Ithmar Capital and NSIA back Trans-Africa pipeline. 21 Dec. https://www.swfinstitute.org/news/50582/ithmar-capital-and-nsia-back-trans-africa-pipeline-3131654614

Sovereign Wealth Fund Institute (2019) Ithmar Capital reveals capital hand. 18 Nov. https://www.swfinstitute.org/news/76108/ithmar-capital-reveals-capital-hand.

State House Abuja (2019) National Economic Council approves additional 250M investment for NSIA. https://statehouse.gov.ng/press-releases/national-economic-council-approves-additional-250m-investment-for-nsia/

The Nigeria Extractive Industries Transparency Initiative (NEITI) (2017) Transfer all oil revenue savings to NSIA – NEITI. https://www.neiti.gov.ng/index.php/media-center/news/369-transfer-all-oil-revenue-savings-to-sovereign-wealth-fund-neiti

US Embassy & Consulate in Nigeria (2020) Joint media release by the Government of Jersey, the Government of the Federal Republic of Nigeria and United States of America Department of Justice. https://ng.usembassy.gov/joint-media-release-by-the-government-of-jersey-the-government-of-the-federal-republic-of-nigeria-and-united-states-of-america-department-of-justice/

US Energy Information Administration (2020) Nigeria https://www.eia.gov/international/analysis/country/NGA

Udo B (2019) Excess crude savings: Why Nigerian govt. must "aggressively" fund NSIA – official. Premium Times, 18 Mar. https://www.premiumtimesng.com/business/business-news/330444-excess-crude-savings-why-nigerian-govt-must-aggressively-fund-nsia-official.html

Ukiwo U, Ukaga U, Ibaba I (2012) Conclusion: Reclaiming politics and reforming governance: Options for sustainable peace and development in the Niger Delta. In: Ukaga O, Ukiwo U, Ibaba I (eds) Natural resources, conflict, and sustainable development: Lessons from the Niger Delta. Routledge, New York, p 152–166

Umoru H (2017) Senate scraps Excess Crude Account, says it's illegal. Vanguard, 8 Nov. https://www.vanguardngr.com/2017/11/senate-scraps-excess-crude-account-says-illegal/

Vanguard (2019) FG commits to improved funding of NSIA. 13 Apr. https://www.vanguardngr.com/2019/04/fg-commits-to-improved-funding-of-nsia/

Watts M (1992) The shock of modernity: petroleum, protest, and fast capitalism in an industrialist society. In: Pred A, Watts M (eds) Reworking Modernity: Capitalisms and Symbolic Discontent. Rutgers UP, New Brunswick, NJ, p 21–64

Wilson Centre (2016) Policy recommendations and options: Harnessing Sovereign Wealth Funds for sustainable development in Africa. The Brown Capital Management Africa Forum. August. https://www.wilsoncenter.org/sites/default/files/media/documents/publication/2016_brown_signature_-_swf_recommendations.pdf

A Less Than Sovereign Wealth Fund: Norway's Government Pension Fund, Global

Jonathon W. Moses

Introduction

Norway's sovereign wealth fund has been admired and copied by states and analysts across the globe. Its admirers hold that the Norwegian Government Pension Fund, Global, or GPFG, is removed from political meddling, and that its politically aloof manner helps Norway avoid the curse of its natural resource abundance. States that follow Norway's lead are advised to quickly funnel their resource wealth offshore, cloister it from political authority, and manage the fund with an eye on economic reward rather than political influence. This depiction, though mostly true, is misleading. If sovereignty refers to the possession of supreme political power, then the GPFG is anything but. The success of Norway's GPFG lies in its responsiveness to *political* authority and interests, not in its autonomy or the technical advice and judgments of its economic experts.

The world's largest sovereign wealth fund is the result of an ongoing series of political struggles, fought out in a unique institutional context. When Norway first discovered oil, it was already a highly functional, stable, and effective social democracy. As a small economy exposed to

J. W. Moses (✉)
Norwegian University of Science and Technology, Trondheim, Norway
e-mail: jonathon.moses@ntnu.no

E. Okpanachi and R. Chowdhari Tremblay (eds.), *The Political Economy of Natural Resource Funds*, International Political Economy Series,
https://doi.org/10.1007/978-3-030-78251-1_8

global markets, Norway developed a form of corporatist governance that responds quickly and effectively to changing global markets (Katzenstein 1985). Its parliamentary system relies heavily on expert committees and the input of producer organizations (representing employers and workers) to generate policies that are derived through consensus and based on expert advice. As a result, Norway enjoys very high levels of public trust, substantial wealth, and remarkably modest levels of income inequality.

From its conceptual origins to its most recent earnings report, the GPFG has changed its nature, direction, and name in response to sundry political pressures. This chapter describes those changes in four parts. At the time of its creation, there was little consensus about the need for a fund, or even what it should look like. Experts advised against establishing a Petroleum Fund and disagreed about its utility. After it was established, and as it matured, the Fund was pulled in different directions: politicians and experts differed on the nature of its investment objectives, and on how the fund should be managed. The second section describes these tug-of-wars, and the frequency with which broad political arguments triumphed over narrow "expert" advice. The third section of this chapter takes a step back to consider the overall operation of the GPFG and what we might learn from its experiences thus far, and in so doing surveys the GPFG's many successes, which were largely the results of those (earlier) political decisions. The fourth part provides a conclusion.

The Origins of the Fund

In February 1958, about a decade before Norway's oil adventure began, the Norwegian Geological Survey wrote to the Ministry of Foreign Affairs, noting that "The chances of finding coal, oil or sulphur on the continental shelf of the Norwegian coast can be discounted" (Gundersen 2007). The experts could not have been more wrong: the first commercially viable fields in Norway were discovered at the close of the next decade (1969), and petroleum production began in earnest in the 1970s.

As the oil began to work its way to the surface, authorities in Norway began to recognize the need to deal with the resulting bounty in foreign exchange. Some first concerns were voiced in an early parliamentary report (*St. meld. nr. 25 (1973–74)*, p. 14), although the word "fund" was not specifically employed. The Ministry of Finance, which authored the report, was concerned that Norway would end up with significant wealth, denominated in US dollars (the oil market is largely a dollar market), and that this wealth would end up offshore, beyond the reach of traditional political guidelines and constraints.

This concern was repeated a decade later, in 1982, when a government committee was established and chaired by a subsequent governor of the Norwegian central bank, Hermod Skånland. This "Pace Committee" [*tempoutvalg*] was assembled to consider how Norway must adapt to its increased oil wealth, and how the Norwegian economy would absorb this wealth. Here, for the first time, the idea of a "buffer" fund was formally introduced. However, the committee advised against adopting a buffer fund, believing it foolish to assume that politicians could keep their fingers out of the cookie jar: "Judging from the attitudes we know in both the political community and the general population at large, it is difficult to imagine that hundreds of billions could be invested abroad, while facing unmet needs at home" (NOU 1983, pp. 27, 90).[1]

To protect the Norwegian economy from this new influx of wealth, the committee recommended that Norway simply slow down the pace of its petroleum activities; hence its name, the "Pace Committee."

Thus, as with the existence of oil itself, Norwegian experts in the Ministry of Finance and at the country's central bank initially doubted the need for a petroleum-related investment fund. Although the "Government Petroleum Fund" was created by the Government Petroleum Fund Act of June 22, 1990 (No. 361), it did not receive its first transfer/deposit until 1996, roughly a quarter-century after the first Norwegian oil flowed to the surface (see Fig. 2). Norwegian doubt over the utility of some sort of buffer fund was fueled by two different types of concerns, both of which were controversial: was a fund really necessary; and, if so, what should it look like?

Did Norway Need a Fund?

Even after Norway began to realize that its petroleum reserves were substantial, it was not clear that the revenues generated offshore should be placed in an international investment fund. Initial resistance to the creation of this fund came in three forms: Norwegian experience with previous funds; express political concerns about how the money might be

[1] Translated in Moses and Letnes (2017a, p. 130).

spent; and disbelief in the need for such a fund, on the basis that current revenues could be better spent on necessities.

Norway's previous experience with investment funds had not set a promising precedent. Its first sovereign wealth fund collapsed shortly after it was created. In 1904, as Norway was still struggling for its independence,[2] its political authorities established the State's Reserve Fund [*Statens Reservefond*] to invest in premium foreign bonds (mostly French, German, and British government bonds). Although this seemed a prudent measure at the time, the inflation generated during and after WWI decimated that Fund, such that it was laid to rest in 1925 (Hylland 2005, p. 182).[3]

The Norwegian government's other main experience with an investment fund was more recent: the Government Pension Fund, Norway [*Statens pensjonsfond Norge*, or SPN]. Remarkably, this fund had an equally spotty record, but for different reasons.[4] Before the arrival of oil, the SPN had been used to finance popular political projects in Norway with relatively poor financial returns (see Lie 2013, 2018). Consequently, there was almost never any surplus left in the SPN.

From both of these experiences, the Norwegian civil service was understandably reluctant to establish a new fund. In particular, they the civil servants were worried that a new fund would be used as a honey-pot, outside traditional political channels, attracting powerful interests. In addition, they worried about losing control over a new and important potential source of government revenue.

[2] Norway was under first Danish, then Swedish, rule until it gained full independence in 1905.

[3] The decision to collapse the Reserve Fund was made by the government, particularly the Ministry of Finance, in 1923. However, doing so required a constitutional amendment. This was secured by a unanimous vote in Parliament in June 1925, when the Fund was liquidated (see Hylland 2005, p. 185).

[4] Originally established in 1967 as *Folketrygdfond*, the Fund was subsequently renamed *Statens pensjonsfond Norge*. Thus, the Norwegian Government Pension Fund has two components: the Government Pension Fund, Global (GPFG) and the Government Pension Fund, Norway (SPN). The SPN is managed separately from the GPFG, by *Folketrygfondet* (the National Insurance Scheme Fund), and its investments are limited to domestic and Scandinavian concerns. It is a key stock holder in many large Norwegian companies, predominantly via the Oslo Stock Exchange. In short, the SPN is much smaller than the GPFG, and functions as a sort of national insurance fund. By the middle of 2018 (Q2), the market value of the SPN was 250bn NOK, while the GPFG (by contrast) was worth 8337bn NOK (see Norwegian Government 2018a, b).

Not only was the civil service largely critical about establishing a fund, there was no consensus about what it should look like. A number of special committees were established to study the utility of a fund, and what it would eventually look like, during a time characterized by substantial political turmoil. The lines of cleavage were many, and cross-cutting: there were differences between the Left and the Right (among political parties), between political elites and civil servants, and even between experts within government.[5] As a result of the shifting political winds, and much indecision and internal infighting, a final decision about the creation of a fund was postponed until 1989, during which time the price of oil had fallen precipitously. Due to this delay, the resulting fund came to life in a context of economic crisis and concern. This brings us to the third important source of resistance.

In 1989–1990, the Norwegian economy was on the ropes, having borrowed great sums of money, against the promise of future oil, to develop its oil resources and to expand its welfare net and social services (see Moses and Letnes 2017a, chapters 4 and 7). This is important to note, for two reasons. First, Norway spent most of its early oil wealth on domestic improvements: it was not until 1996 that the government was able to tuck away any surplus in an oil fund.[6] Second, and relatedly, in 1989–1990, very few people thought the fund would amount too much, especially given Norway's previous experience with funds. Because there was little faith in the potential of the fund, as there had been in the existence of oil offshore to begin with, it was possible to reach a consensus on the particular nature and contours of that fund. In short, the stakes did not seem to be very significant at the time.

What Should It Look Like?

The second, and related, issue of contention was over what the fund should look like. Throughout the 1980s, there were a series of skirmishes

[5] The details of these disputes are outside the scope of this chapter. See Lie (2013, 2018) for a fuller description and references to the different committee reports and their conclusions.

[6] About ten percent of Norway's public expenditures is paid for by petroleum revenues, which has been the case since oil was first discovered. Before 1994, the money came in the form of taxes and sales of petroleum and went directly into the national budget (Bye et al. 1994, p. 36). After 1994, this money came from returns on investments made by the Petroleum Fund—subsequently the GPFG (Qvigstad 2012, pp. 79–80).

about how this money might be used, even if many of the commentators were skeptical that a fund would ever amount to much. For example, proposals came from the right to use the money as foreign aid[7] and from the left to export social democratic goals to a global community.[8]

A parliamentary report in the early 1980s from the Ministry of Petroleum and Energy (MPE) suggested that the oil revenues might be used to build domestic infrastructure, industry, and growth capacity (Lie 2018, p. 292). This prompted the creation of a new expert committee, including representatives from both the Ministry of Finance (MoF) and the MPE, which soon found itself divided.[9] A majority in this committee, captured by the MoF, suggested that the money should be deposited in a government deposit account or a foreign currency deposit account at the central bank, governed by ordinary fiscal policy controls. The minority opinion, made up of experts from the MPE, proposed a more traditional "fund," with both savings and buffering functions, and where the investments were to be made in low-risk, fixed-income securities or shares in large companies. This (minority) group also entertained the possibility of using the money for domestic investments, helping foreign companies invest in Norway, and/or even to facilitate Norwegian takeovers of foreign companies (Lie 2018, p. 292).

The end result of this tug-of-war was a grand compromise that lay the foundations for today's GPFG. Indeed, the compromise solution ended up being very similar in design to the original "buffer" fund that was first entertained in the Pace Committee report. Although the MoF originally opposed a fund, it could agree to one located offshore. Hence, the compromise ensured that all revenues from petroleum activities would be transferred directly abroad, to avoid flooding the Norwegian economy. At the outset, these investments were limited to fixed securities.

One key point of contention concerned how the money would be subsequently repatriated to the Norwegian economy from these foreign

[7] This was one of the main conclusions from the Conservative Party's oil committee report "Oljevirksomheten og Norges fremtid" in May 1981 (see, e.g., Skredderberget 2015, p. 54).

[8] Per Kleppe, the Labour Party Finance Minister in the mid-1970s, recognized that the money needed to be invested abroad rather than used to increase wages or consumption, but he hoped the money would be used to build a more egalitarian, greener, and gender-neutral world (see Hanisch 1999, p. 24).

[9] For details and references, see Lie (2013, pp. 331–332).

investments. This too was solved by compromise: these revenues could only be invested in foreign markets, and any spending of the return from the capital invested had to be approved by the Parliament, as part of the regular government budgetary process. Today we refer to this process as the "budgetary rule" [*handlingsregel*], which was initially discussed in the 1982 Pace Committee (see Moses and Letnes 2017a, p. 132).

When the Pace Committee first recognized the need for such a rule, it pointed to the need to ensure that the rate at which the oil money entered the Norwegian economy would be both consistent (stable) and economically prudent. While the caveat was clear, the particular details of this rule of thumb were not established until 2001, when the Finance Ministry published a white paper (*St. meld. nr. 29 (2000–2001)*).[10] This report noted that the return should be determined by the expected real rate of return from the fund (*St. meld. nr. 29 (2000–2001)*, p. 9); however, what this actually means has been a matter of varying political interpretation. In 2001, the government, with broad political support in Parliament, suggested a 4% threshold, believing that this corresponded with the long-term return on the fund's investments. In other words, the government could use as much as 4% of the annual projected return from the fund to pay for its budgetary expenditures. In 2017, another government reduced that threshold to 3%, in light of global economic conditions, arguing that the long-term return of the fund was now likely to be lower.

In this way, Norway's oil money can enter into the economy at a steady and predictable rate in contrast to the volatile price of oil, allowing governments to plan accordingly. At the same time, the system is flexible enough to allow authorities access to more money in times of economic crisis (Moses and Letnes 2017b). Hence, the GPFG acts like both a savings and a Stabilization Fund: it saves money for future need, but can also be used to stabilize the broader Norwegian economy from the volatility associated with global petroleum prices, and the threat that the resulting wealth brings in the form of Dutch Disease.

Hence, by 2001, the main pillars of the Norwegian sovereign wealth fund had been established, and these pillars have remained in place, despite several smaller changes in subsequent years. In particular, there is now broad consensus that:

[10] See Moses and Letnes (2017a, pp. 131–134) and Moses (2020) for further details.

- All revenue from petroleum activities should go directly to the offshore fund;
- All investments in the fund should be made abroad; and
- All spending of the fund (or its return) must be approved by the government, as part of the government budget balancing procedure.

Although dissent about these pillars has surfaced on occasion, such as the suggestion that more money should be spent in Norway on productive investments, there has generally been a strong and broad consensus about them. As the Fund has grown and matured, the political discussion has focused, instead, on the appropriate make-up of its investments and the management structure of the Fund. These controversies are the subject of the next section.

An Evolving Fund

Money first began to flow, quickly, into the Petroleum Fund in 1996. As the Fund grew, so did political discussions about how to use this money in an appropriate and politically responsible manner. These discussions revolve around the particular make-up of the Fund and how it should be managed. Here, as during its creation, a cleavage has developed between those experts responsible for managing the Fund, and those who hoped to use this money to secure larger political objectives.

Make-Up

When it was first formally established, in 1990, the Fund was thought of as a relatively small, back-room operation at Norges Bank (NB), the central bank of Norway. When the first deposits were made in 1996, NB was tasked with investing the money in the same way as it did with the nation's foreign exchange reserves. This meant a very conservative, safe return on the people's money in government bonds over a relatively short temporal horizon, in markets that were familiar to the central bank. This staid conservative approach did not last long.

It took only one year after the first deposits before the Petroleum Fund's portfolio were broadened to include equities. In 1997, NB was instructed to quickly and quietly place 40% of the Fund's holdings in private equities. In doing so, it was made clear that the central bank should *not* use its ownership actively, such as voting its shares, as it simply did not have the competency or the manpower to do so satisfactorily or to risk unnecessary controversy (Skredderberget 2015, p. 79). As the Fund would eventually require a more active management team, the Ministry of Finance established a new department at the central bank, first called *Norges Bank Kapitalforvaltning* and subsequently renamed Norges Bank Investment Management, or NBIM, at the start of 2008.

Over the years, NBIM grew the Fund within guidelines set by the Norwegian Parliament, or Storting. The changes include significant shifts in the nature of the portfolio; for instance, its equity holdings were allowed to expand into developing markets and grow in size as a share of the total. In 2000, five emerging markets were added to the fund's benchmark equity index; in 2002, corporate and securitized bonds were added to the Fund's benchmark fixed-income index; in 2008, all emerging markets were included in the reference index for equities; and in 2012, European holdings were reduced, while emerging market holdings were again expanded.

In 2007, the Storting pushed to broaden the Fund's portfolio even more, increasing its equity share from 40 to 60% while also adding small-cap companies to the benchmark portfolio. Although it was not planned in advance, the rapid expansion in equity shares, planned a year before the financial crisis, became part of NBIM's response strategy to the 2008 financial crisis (see below). Finally, in 2008, the Fund began to include real estate investments, at first limited to 5% of total assets; the first real estate investment was not made until 2011.

The current guidelines for the Fund reflect this altered investment landscape: up to 70% of the Fund can be invested in equities; as much as 7% can be invested in real estate; and as much as 30% can be used on fixed-income investments.[11] Of all these changes over the years, the greatest and perhaps best known of these came in 2004, when a Council on Ethics was established, developing guidelines that were implemented in 2006.[12] The initial pushback to this Council, from both Norges Bank and the Ministry of Finance, was significant. The country's central bank, Norges Bank, was explicit in its condemnation, warning that the proposed ethical guidelines "could impose significant costs" and expose the Fund to

[11] See https://www.nbim.no/en/the-fund/holdings/holdings-as-at-31.12.2018/?fullsize=true for a detailed overview (and interactive map) of NBIM's holdings.

[12] See Moses and Letnes (2017a, pp. 136 and 215–219) and Moses (2020) for further details.

increased and unnecessary risks (Skredderberget 2015, p. 124). However, the perceived hypocrisy of Norwegian policy eventually pushed political officials to adopt stronger ethical guidelines.

The backstory to the Ethics Council is fascinating, but outside the scope of this article.[13] Much of the initial pressure for ethical guidelines came from the radical left in Parliament, especially Øystein Djupedal, the Socialist Left (SV) representative in the Storting's Finance Committee. Then, in 1997, Oslo hosted an international delegation to discuss a global prohibition on land mines (the same year that Jody Williams received the Nobel Prize for her efforts in this area), and the Norwegian government became a strong and vocal advocate of the ban. In this context, the SV's revelation in 2001 that the Petroleum Fund had been investing in Singapore Technologies, a company that produced land mines, was embarrassing: Norway appeared to be fighting for a global ban on land mines while profiting from investments in firms that produced land mines. This revelation eventually forced the Petroleum Fund to sell its shares in Singapore Technologies, even though the explicit public justification was based on a concern about breaking international law rather than on ethical concerns.

At the same time, the Norwegian newspaper *Dagbladet* ran a series of investigative reports by the journalist Thomas Ergo on how Fund investments were being made in a number of questionable areas including weapons production, tobacco companies, and firms that used child labor or even slave-like working conditions. As political pressure mounted, the Centre-Right Bondevik government gave way and followed SV's lead. In the fall of 2002, a committee was established (*Graver utvalget*), and its recommended ethical guidelines and the Petroleum Fund's Advisory Council on Ethics was established in the fall of 2004.

Since 2004, the Fund has been required to follow strict ethical guidelines, with new ethical measures being added subsequently (see https://etikkradet.no/en/). Today, the Council on Ethics is an independent body made up of five members, with a staff of eight. These members, and the chair and vice-chair from them, are formally chosen by the MoF, on advice from the NBIM, and sit for a period of four years. The required competencies to sit on the Council are only vaguely expressed: members should "have that competence that is required in order to execute their

[13] For readers of Norwegian, see Skredderberget (2015, p. 124ff).

assignment" (NBIM 2016b, §4). The Council on Ethics makes public recommendations, on the basis of guidelines determined by the MoF for Norges Bank (see Ministry of Finance 2017a), which help to establish whether a given company should be excluded from the GPFG or placed under observation.

The nature of today's GPFG investments is quite different from the way that the original Fund was conceived. The most significant difference is the very public manner by which these investment decisions are made, in light of public discussions and transparent ethical standards. The Fund is actively managed, with significant political (ethical) influences, and generates great rewards.

Management

The same sort of unexpected development can be seen with regard to how the Fund is managed. The discussion here is a frame in terms of big picture issues and little picture issues. The former concerns allocating political responsibility: how can we ensure that the decisions being made reflect the people's interests, and not the narrow interests and perspectives of experts, when these differ? The latter focuses on how decisions were made on the ground and about where investments should be made: to what extent are these decisions being made in-house or made by hired market actors, and who is calling the shots? Answers to these questions are interesting in that responsibility for these decisions has changed over time, and the two issues are now interrelated as new management structures continue to be considered and proposed.

With regard to the big picture, the management of the Fund involves three sets of political actors who are frequently at odds: elected politicians, civil servants at the MoF, and officials at Norges Bank. Formally, the main framework for the Fund was established by the Storting, in the Government Pension Fund Act of 2006.[14] The 2006 Act formally changed the name of the fund from a Petroleum Fund to the "Government Pension Fund, Global" (GPFG). The initial motivation for Norway's fund was to establish savings that could help the country transition to a post-oil economy once the petroleum ran dry. To date, there has never been an

[14] English Summary: https://www.nbim.no/en/organisation/governance-model/government-pension-fund-act/; Norwegian version: https://lovdata.no/dokument/NL/lov/2005-12-21-123?q=lov+om+statens+pensjonsfond.

explicit political decision as to how the money in the fund should be used in the future; the reference to pensions was simply a convenient way for politicians to signal the need to save for the future.

While the Norwegian Parliament has always exerted significant influence over the make-up and management of the Fund, experts in the central bank and the MoF have substantially pushed back against it.[15] What is remarkable is that the interests of politicians from varying ideological blocks have frequently united in opposition to the advice provided by experts.[16] This difference in approach is best explained by the responsiveness of Norwegian elected officials to public opinion, whereas technical experts are mostly isolated from this sort of pressure.

The Ministry of Finance has overall responsibility for the Fund and has issued guidelines in its Management Mandate[17] for the GPFG. This is interesting if only because the MoF initially advised against creating an oil fund. As noted earlier, the MoF doubted that this type of fund would be able to secure sufficient capital and they were afraid that politicians would be tempted to use the money for their own purposes; they have, thus far, been proven wrong on both accounts.[18] Even so, the Finance

[15] Of course, this is not always the case. Mainstream politicians initially lined up with experts at the MoF and NB to oppose the introduction of ethical guidelines: it was only the concerted effort of radical-left politicians that won the day, and only later did the rest of the elites followed in line when they saw which way the political winds were blowing.

[16] Before the Conservative Finance Minister, Arne Skauge, made the fateful decision to create a Petroleum Fund, he called the Labour Party's financial spokesman (in opposition), Sigbjørn Johnsen, to ensure his support. In effect, the political elite were united, in opposition to the experts at the MoF, to create a Fund (see Skredderberget 2015, pp. 70–71). Later, in 1997, the decision to expand into equities (40%) was made by Jens Stoltenberg, after his Labour Party had lost power in parliamentary elections, but while he still sat as Finance Minister. Stoltenberg pushed this expansion through, even though the incoming Finance Minister from the Center Party, Gudmund Restad, had been skeptical. Once in office, however, Restad continued Stoltenberg's policy (see Skredderberget 2015, p. 84).

[17] https://www.nbim.no/en/organisation/governance-model/management-mandate/.

[18] One can only speculate as to why Norwegian politicians appear to be more responsible than politicians from other countries. Norway's unique political and institutional context is surely relevant to any explanation. For instance, political elites in Norway constitute a small and homogeneous group that are often in contact with one another. As a result, it may be easier to defend a broad consensus among themselves than it is to break out with a populist proposal to raid the fund, for example.

Department is tasked with providing an annual report to the Parliament in order to inform the people's representatives.[19]

Finally, the Executive Board at Norges Bank (NB) is tasked with managing the Fund. The eight members of the Executive Board are appointed by the Government, with the governor of Norges Bank sitting as its Chair. Five of the eight members are not from the staff at Norges Bank, but are picked from the ranks of the academic and private sectors (Norges Bank 2019). This Executive Board meets 16–17 times a year to manage the central bank's broader activities (including NBIM), but it delegates operational management of the fund to the CEO of NBIM, who assembles a leader group that includes a Deputy CEO and four Chief Investment Officers responsible for the fund's strategies (NBIM 2019d; see Fig. 1).

This division of responsibility is rather curious, in that the central bank first resisted the job. The initial decision to place the Fund at NB was made because politicians recognized NB's experience with managing the country's foreign reserves, and because the MoF hoped to create some distance between the Fund and elected politicians. NB's original resistance was based on a concern about expanding into new activities, and the effect this would have on its reputation, as NB had neither the expertise nor the culture to manage a large equity or real estate portfolio.

With regard to the smaller picture, it is important to realize that Norges Bank was not alone in lacking competency for managing a fund of this nature, as this sort of expertise was also underdeveloped at the Finance Department. When Norway decided to wade deeper into equity markets in 1996, it chose four international index managers (Barclays, State Street, Bankers Trust, and Gartmore) to allow a slow and careful entry into the market in a way that would not cause ripples, given that the companies involved were not made public at the time. This allowed them to develop their position by means of internal crossing; in other words, the companies did not go on the market buying positions in the name of the Fund, but purchased from existing clients, internally, in each of the four companies.

Once the required amount of equities was secured, the Fund began to develop its own management competence. In 1998, it conducted a search for help in actively managing the Fund and received responses from 260

[19] Remarkably, and despite its reputation for transparency, there was no external review of how the GPFG was managed before 2006.

companies, only three of which were eventually chosen to help *Norges Banks Kapitalforvaltning* become more proficient in active management (Skredderberget 2015, p. 93). Still, it was not until 2006, sixteen years after its formation, that the Finance Department established a Department for Wealth Management [*Avdeling for formuesforvaltning*], and secured an external review of the management practices of the Fund. Over time, these competencies grew spectacularly, with offices opening up in London, New York, Shanghai, and Singapore. Today, the NBIM employs over 600 people in a number of specialized areas of competence. Almost half of these employees (42%) work in one of the overseas offices, and a substantial share is now working in real estate (22%).[20]

Like the specific make-up of the Fund, the management structure of the GPFG—not to mention its name—has changed substantially over the years, although there is still no consensus about the best way to manage the Fund. As the GPFG becomes more successful, the management issues become more critical, and the rewards associated with management responsibility grow.

Operational Overview

Up to this point, I have been painting a picture of continuous skirmishes between political and technical elites. It is a dynamic picture, frequently changing in response to conflicting pressures. This section provides a static snapshot of the guidelines by which the GPFG operates today. As explained above, this operation came into being because of Norway's unique history and context. This tension between technical and political elites in Norway is not unique to the management of the Fund, but is a typical feature of Norwegian democracy in practice. The second part of this section elaborates upon this unique context, and hints at how it points to the potential for future skirmishes over the GPFG. The third and closing part of this section considers the many successes that have resulted from the success of Norwegian policymakers at creating a popular and wealthy Fund that is responsive to the public's hopes and concerns.

[20] See NBIM (2018, p. 130).

Operational Snapshot

The previous two sections have described a dynamic process in which institutions and objectives result from conflicting interests and opinions. However, outlining this process may complicate a view of the main objectives and mandates that characterize today's GPFG. This section provides an overview, by briefly describing some of the Fund's most important operational characteristics.

The most important of these operational characteristics is the fact that the GPFG is *not* a separate and autonomous legal entity. Although it is well known that operational management of the GPFG is delegated to the NBIM—while the real estate management unit is organized as a separate sub-entity, Norges Bank Real Estate Management, or NBREM (see Fig. 1)—most people do not realize that the NBIM is not autonomous: it answers to Norges Bank, and Norges Bank (in turn) is directed by a management mandate set by the Ministry of Finance. This chain of command is clearly evident in Fig. 1.

In addition, it is important to know the guidelines that determine how deposits and withdrawals can be made in the Fund. In both cases, the general processes are outlined in the Government Pension Fund Act (NBIM 2015). There are three sources of income deposited in the GPFG: the net cash flow from petroleum activities, which is transferred from the central government budget; the net results of financial transactions associated with petroleum activities; and the return on the Fund's capital (NBIM 2015, §3).[21] As noted earlier, the central government is only allowed to tap into this fund as part of its normal budgetary procedures, pursuant to a resolution by the Norwegian Parliament: it does so in a way that covers what is called "the oil-adjusted budget deficit," using the "budgetary rule" [*handlingsregel*]. It does this on an annual basis, and almost always within the limits set by the budgetary rule, which is relaxed only during periods of recession. These operational characteristics are important, as they set the formal guidelines that govern the GPFG. However, the governance of the GPFG is also affected by national contexts and historical experiences.

[21] When deciding how to invest this money, NBIM is guided by a *management objective* "to maximise the return on the Fund as measured in international currency, within a moderate level of risk. Within this overarching financial objective, the Fund shall be responsibly managed" (Ministry of Finance 2017b, p. 3).

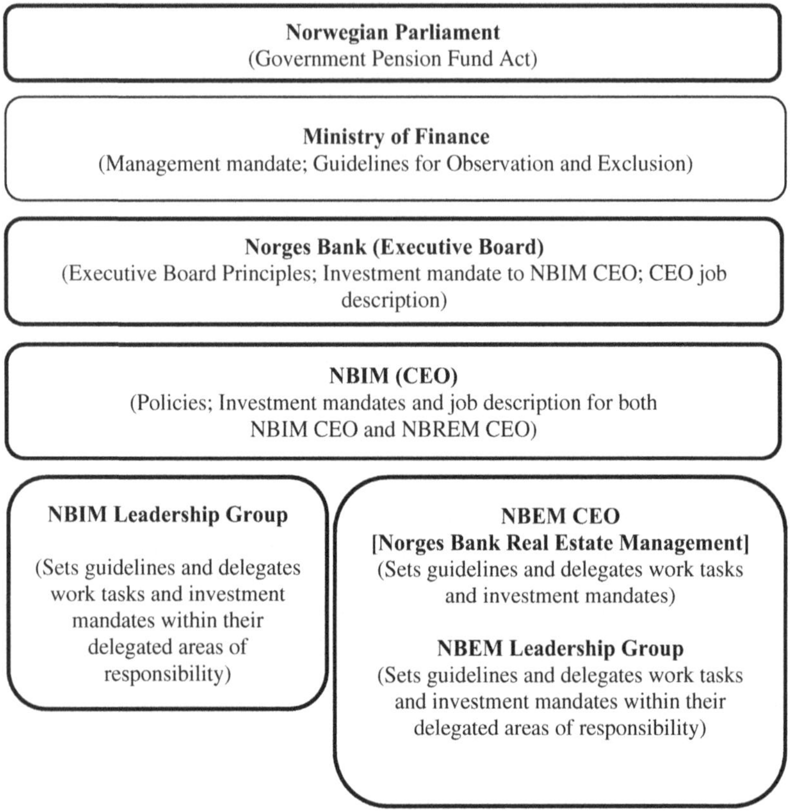

Fig. 1 Operational Structure of the GPFG

Unique Context

As mentioned in the introduction, Norway's stable democratic history and its long-term reliance on a number of powerful and independent interests and institutions play important roles in explaining the country's success in managing its petroleum fortunes. In addition to this broader political and institutional framework, Norway's disappointing history with prior funds influenced the path of its subsequent development. Norway's initial reluctance to establish a fund allowed political authorities more time to spend their early oil revenues to pay down the national debt and to build up the welfare state. Because experts had been wrong in the past about the utility of such funds and about the existence of oil, Norwegian policymakers were able to maintain a remarkably sober attitude with respect to what a future fund might, or should, look like.

This is the context in which Norway developed its GPFG. The particular nature of the GPFG was determined by the give and take of democratic compromise. There were several overlapping cleavage lines in this struggle to create the GPFG, but the most enduring cleavages were between elected officials of varying political stripes and economic experts, and between the experts representing the most influential institutions (MPE, MoF, NB, NBIM). Although the main political parties experienced ideological differences over the establishment of the fund, as described in the first section, these differences quickly dissolved when a consensus was developed for creating the fund. Since that time, the main question has been how to ensure that the GPFG reflects the values of the Norwegian people while still delivering a healthy reward.

This is noteworthy as it is too easy to look at Norway's success and hope that by importing similar institutions, a country might hope to mimic Norway's achievements. Norway does not always follow the expert's rulebook, and its success has been contingent on its ability to improvise and follow public opinion. For example, Norway helped to create the International Forum of Sovereign Wealth Funds (IFSWF), and it regularly demonstrates its adherence to the Santiago Principles (see, e.g., Ministry of Finance 2017b). But Norway is not actually a member of the IFSWF,[22] and as we have seen, the GPFG is not a legally independent entity. This is important, as it is *the people of Norway*, as voiced through their Parliament, and as reflected in the Ministry of Finance, that set the objectives for the GPFG, rather than a technocratic board of experts.

As the Fund has grown, and as the NBIM has taken on a broader set of responsibilities, the management structure has been subject to change and debate. One example of this is the fact that the GPFG's Council on Ethics now reports directly to the central bank, on the basis of ethical guidelines determined by the Ministry of Finance; before 2015, the Council

[22] Norway's reluctance is apparently in response to the way that the IFSWF has developed, in that it is not working hard enough "to encourage and ensure transparency about the management of sovereign wealth funds, including objectives, governance framework, investments and risk management" (see IFSWF 2018, p. 46).

reported directly to the MoF. Another example is the continued disagreement about where management responsibility for the Fund should lay. As illustrated in Fig. 1, the NBIM currently answers to Norway's central bank, Norges Bank, which in turn is responsible to an Executive Board,[23] appointed by the government, with broad-ranging political and economic expertise. This means that the NBIM responds to a body that has limited expertise in the area of financial planning. In recent years, there has been some discussion about splitting up the GPFG into different groups such as securities, equities, and real estate, and providing each with its own board of directors, specifically suited for the unique needs within each investment category (e.g., Skredderberget 2015, p. 171). Still, a 2017 government commission (NOU 2017, p. 13) recommended that the GPFG should be kept together as a single unit but be given more autonomy from the central bank. Although this discussion has not developed along political/ideological lines as of yet, a new ideological cleavage may be developing over how the Fund should be spent (see below).

In short, the experts are again at odds with the political consensus with their recommendation of greater autonomy for the NBIM. Doing so would make the NBIM an autonomous legal entity, but it would also distance the NBIM from the sort of democratic pressures that have made it so successful in the past. As of this writing, there have not been any changes made to the existing management regime at NBIM, but it is easy to predict a new line of cleavage developing between elected representatives in Parliament and the expert advice provided by NOU (2017, 13).

Dynamism

In light of Norway's experience, and in contrast to much expert advice, the success and dynamism of the GPFG can be tied to its responsiveness to political influences. The scope of this dynamism is wide-ranging, but due to space reasons, this section discusses four different types of success achieved by the GPFG: its size, its market influence, its capacity to stabilize the broader Norwegian economy, and its overall performance as an investment fund.

[23] https://www.norges-bank.no/en/topics/about/Organisation/The-Executive-Board/. There is also a Supervisory Board, which supervises the Executive Board to ensure that the Bank's activities are conducted in accordance with legislation, agreements, decisions, and other regulatory frameworks.

Size

Despite political involvement and ethical constraints, the GPFG has been remarkably successful. As shown in Table 1, the GPFG is today the world's largest sovereign wealth fund, a position it has held for several years. After passing the 1 trillion (US) dollar threshold in 2017 (NBIM 2017), it remains a huge and healthy fund, with investments representing roughly 1.3% of all the world's listed companies (Moses and Letnes 2017a, 135). As of September 2019, there have been reports of record high growth for the GPFG in 2019 (Bjørnestad 2019).

Figure 2 shows the growth of this fund, over time, both in NOK terms and as a percentage of GDP. It traces the Fund from its modest beginning in 1996 to its phenomenal growth in the twenty-first century, even during the financial crisis of 2008. The Fund is now growing from returns on investment as much as from new deposits from petroleum activity in Norway, which remains strong, despite some decline over time.

Table 1 Largest SWFs by Assets under Management

	Country	*Name*	*Assets*	*Origin*
1	Norway	Government Pension Fund, Global	1074,6	Oil
2	China	China Investment Corporation	941,4	Non-Commodity
3	UAE Abu Dhabi	Abu Dhabi Investment Authority	683	Oil
4	Kuwait	Kuwait Investment Authority	592	Oil
5	China-Hong Kong	Hong Kong Monetary Authority Investment Portfolio	522,6	Non-Commodity
6	Saudi Arabia	SAMA Foreign Holdings	515,6	Oil
7	China	SAFE Investment Company	441	Non Commodity
8	Singapore	Government of Singapore Investment Corporation	390	Non-Commodity
9	Singapore	Temasek Holdings	375	Non-Commodity
10	Saudi Arabia	Public Investment Fund	360	Oil

Note Assets listed in Billions USD
Source SWFI (2018)

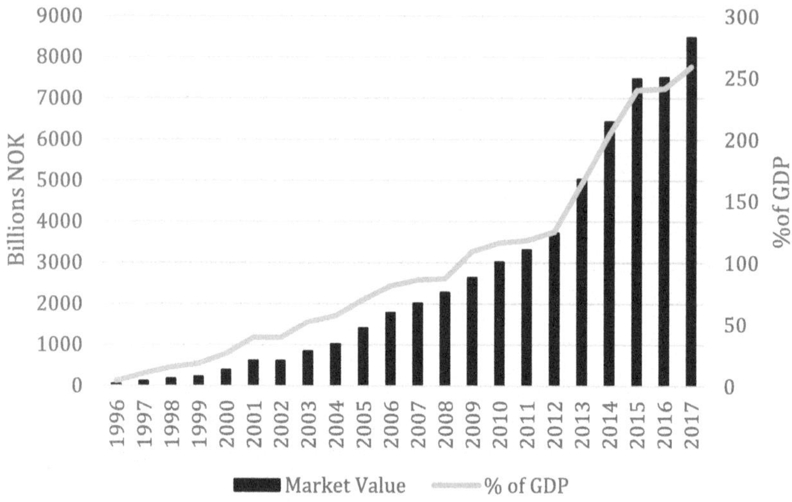

Fig. 2 GPFG, 1996–2017 (*Source* NPD [2018])

Market Influence

The second important result of the GPFG is its capacity to influence market behavior. As pointed out above, this is not something that the Fund, or Norwegian policymakers, set out to accomplish, but the NBIM now seems to embrace its leadership role. This market influence is through the ethical guidelines themselves and through the NBIM's role as an active owner.

First, the Council on Ethics sends important signals to investment markets about what does and does not constitute ethical behavior. These guidelines include product-based and conduct-based exclusion criteria. Product-based criteria can be used to exclude firms that produce tobacco, coal, or certain types of weapons, while conduct-based criteria search out gross corruption, human rights violations, environmental damage, and/or unacceptably high greenhouse gas emissions (Council on Ethics 2017, p. 7; see also Council on Ethics 2019). Given the size of the GPFG, and the publicization of the Council's recommendation (Council on Ethics n.d.), prospective firms, and other investors pay attention to the Council's recommendations. Then, when the NBIM makes its decision against the backdrop of these recommendations, the list of excluded companies is made public (see NBIM 2019a).[24]

In 2019, a new cleavage line developed over the degree to which the Fund should move in a greener direction. At the start of the year, the Parliament decreed that the GPFG could invest as much as 120bn NOK in renewable energy projects that are not listed on stock markets. This sort of investment was not to exceed 2% of the fund's total value (Haakonsen and Sandvik 2019). In March, the NBIM announced that it would divest from 134 companies that explore for oil and gas, worth almost $8 billion USD (Carrington 2019). Then, in November, the leader of the opposition Labour Party suggested more political influence over Fund investments, to ensure that the Fund was encouraging a switch to renewable energy. This, in turn was followed by a comment from a high-ranking member of the Socialist Left (SV) party, who suggested for the first time that some of the Fund should be invested in Norway (Haakonsen and Sandvik 2019). All of this triggered a strong rebuttal from the Prime Minster, Erna Solberg, who responded that "such thinking is very dangerous for Norway as a country" (Hågensen and Liahagen 2019). This example demonstrates how growing political tension over climate change is playing out in domestic political discussions about how the Fund should be investing. Given the size and influence of the GPFG, this sort of domestic political debate will likely influence broader market developments about the need to encourage more investments in renewable energy.

Finally, the NBIM also employs its voting rights, and its dialogue with the boards of companies in which it invests, to improve governance and performance, and to encourage more sustainable business practices. As much of this lobbying and coaching is done in a public manner, and because the NBIM's voting record is made publicly available (see NBIM 2016a), NBIM is able to pressure companies to behave in a more ethical and sustainable manner.[25]

[24] In 2019, for example, three companies were held out for special mention: Texwinca Holdings Co was excluded for serious or systematic human rights violations, while Evergy Inc. and Washington H. Soul Pattinson & Co Ltd. were excluded based on an assessment of the product-based coal criterion (NBIM 2019b). A full list of the excluded companies can be found at https://www.nbim.no/en/the-fund/responsible-investment/exclusion-of-companies/.

[25] See Moses (2020) for more details about how the NBIM influences markets.

Stabilization

The GPFG also plays a very important role in stabilizing the Norwegian economy during periods of falling and/or fluctuating commodity prices. The success of this strategy was clearly evident during the drop in petroleum prices in 2014. That fall in prices meant that a smaller amount of money was transferred into the Fund from petroleum sources. For many countries reliant on petroleum, this price fall had a dramatic effect: the capacity of the economy to adopt to changing market conditions was undermined by the fall in government revenues associated with declining petroleum prices. Because of the GPFG, Norway enjoyed a much broader scope of policy autonomy, since most of the new capital entering the GPFG comes in the form of returns from previous investments rather than from petroleum. These returns continue to flow into the GPFG investment kitty, even when the price of oil/gas falls dramatically. The government, then, was able to access the return on that fund to invest in counter-cyclical measures in those areas of the economy most affected by the fall in petroleum prices. This allowed the government to adapt quickly to the changes in market conditions (see Moses and Letnes 2017b).

Performance

Finally, we should not ignore the remarkable investment prowess of the NBIM. One of the reasons that the GPFG has been able to grow so large is its successful investment strategy. By focusing on long-term gains, and by training up an internal staff which is able to exploit the short-term myopia of many market actors, not to mention large doses of luck involved as well, the GPFG has been notably successful.

The success of that strategy is perhaps most evident in how the GPFG was able to rebalance during the 2008 financial crisis: it adjusted its portfolio at a time when the market was flooded with shares, keeping in mind that NBIM was tasked with increasing the equity share of the portfolio from 40 to 60% in 2007.[26] While the GPFG suffered a negative

[26] When the stock market falls, by, say, 20%, this means that the value of NBIM's holding of stocks falls by 20%, even as the NBIM guidelines require it to have 60% of its holdings in equities. This means that NBIM has to buy more stocks to meet its guidelines, though the price of those stocks is falling. During the financial crisis, NBIM was buying a great many stocks at rock bottom prices, knowing it would be able to hold on to them over the long run and enjoy their subsequent rise in value.

return of 23.3% in 2008, the Fund grew by a remarkable 25.6% in 2009 (Skredderberget 2015, pp. 156, 158). Overall, in the period of 1998 to 2018 as a whole, the annual return on the GPF was 5.5%; and the annual net real return, after deductions for inflation and management costs, was 3.6% (NBIM 2018, p. 9). Finally, in the first quarter of 2019, the GPFG enjoyed its highest return ever: 9.1%, or 738 billion kroner (NBIM 2019c), a praiseworthy accomplishment.

Conclusion

We like to think of sovereign wealth funds as apolitical, technical marvels of portfolio management, steered by market experts. We are told that the state can benefit most by adopting a hands-off approach: we should create an autonomous investment entity and simply let the expert investors do their job. The Norwegian experience is mostly contrary to this common view. As we have seen, Norway's sovereign wealth fund is the product of significant procrastination, political infighting, and struggles, and the advice of experts has frequently been proved wrong, or at least short-sighted.

The most important lesson to be learned from the Norwegian experience is that the revenues from petroleum belong to the people of Norway and these revenues must be spent or invested in accordance with those people's needs. In the beginning, the country chose to build out its welfare state and infrastructure, so that the people could benefit directly from the oil revenues. Once this was done, and additional oil revenues brought with them the threat of the dreaded Dutch Disease, the people's representatives chose to invest the money in a way that would not undermine the competitiveness of Norway's non-oil export economy.

In developing this external investment fund, the people of Norway demanded a say in how this money would be invested, and they have been actively involved through their representatives in Parliament. As the people's representative, NBIM is playing an increasingly activist role. Its ownership decisions have a marked influence on the investment decisions of other public entities in Norway, the international investment community, and the individual companies in which it does or does not invest. It is important to note that this activism does have limits, both in terms of what constitutes ethical behavior and in terms of the size of its holdings in any particular entity, but these limits are continually exposed to political discussion and open to change.

Herein lies GPFG's greatest asset. Its strength does not lie in its size, power, and/or technical expertise, although all these characteristics are impressive in themselves. Norway's sovereign wealth fund acts less like a leviathan and more like a trustee. The real strength of the GPFG lies in its capacity to reflect the changing hopes and aspirations of the Norwegian people, whose financial future it holds in its ledgers.

Acknowledgements Paper presented to the workshop on "The Political Economy of Natural Resource Funds" held at the University of Victoria, 27–28 September 2019. I would like to thank participants at that workshop, and especially Eyene Okpanachi, for their helpful comments.

References

Bjørnestad S (2019) Statsbudsjettet for 2020 drukner i oljepenger. Aftenposten, 6 Sept

Bye T, Cappelen Å, Eika T, Gjelsvik E, Olsen Ø (1994) Noen konsekvenser av petroleumsvirksomheten for norsk økonomi. Statistisk sentralbyrå, rapporter, 94/1. SSB, Oslo

Carrington D (2019) "Historic breakthrough": Norway's giant oil fund dives into renewables. The Guardian, 5 Apr. https://www.theguardian.com/environment/2019/apr/05/historic-breakthrough-norways-giant-oil-fund-dives-into-renewables. Accessed 11 Nov 2019

Council on Ethics (n.d.) Recommendations. Council on Ethics for the Norwegian Government Pension Fund Global. https://etikkradet.no/recommendations/. Accessed 4 Sept 2019

Council on Ethics (2017) Annual report 2017. Council on Ethics for the Norwegian Government Pension Fund Global. https://nettsteder.regjeringen.no/etikkradet3/files/2018/03/Etikkradet_arsmelding_2017_eng_UU.pdf. Accessed 4 Sept 2019

Council on Ethics (2019) Council's activities. Council on Ethics for the Norwegian Government Pension Fund Global. https://etikkradet.no/councils-activities/. Accessed 4 Sept 2019

Gundersen I (2007) Oil veterans look back. Norwegian Petroleum Directorate, 2 Mar. http://www.npd.no/en/news/News/2007/Oil-veterans-look-back-/. Accessed 19 Mar 2015

Haakonsen A, Sandvik MD (2019). Presser på for økt styring. Klassekampen, 11 Nov, p 4–5

Hågensen FO, Liahagen JF (2019) Kjempefarlig for Norge som nasjon. https://www.tv2.no/a/10973995/. Accessed 11 Nov 2019

Hanisch TJ (1999) Overgang fra en avventende til en aktiv oljepolitikk, 1965–72. In: Hanisch TJ, Nerheim G (eds) Norsk Oljehistorie. Fra vantro til overmot? Norsk Petroleumsforening, Oslo, p 127–186

Hylland A (2005) Statens reservefond 1904–25. Et forsøk på å binde politisk handlefrihet? Penger og kreditt, nr. 3

IFSWF (2018) The origin of Santiago principles. Experiences from the past; guidance for the future. https://www.ifswf.org/sites/default/files/IFSWF_Santiago_Principles_book.pdf. Accessed 11 Nov 2019

Katzenstein P (1985) Small states in world markets. Cornell UP, Ithaca, NY

Lie E (2013) Bukken og pengesekken. Nytt Norsk Tidsskrift 30(4):323–336

Lie E (2018) Learning by failing: The origins of the Norwegian oil fund. Scandinavian Journal of History 43(2):284–299

Ministry of Finance (2017a) Guidelines for observation and exclusion of companies from the Government Pension Fund Global. https://www.regjeringen.no/globalassets/upload/fin/statens-pensjonsfond/formelt-grunnlag/guidelines-for-observation-and-exclusion-from-the-gpfg---17.2.2017.pdf. Accessed 5 Oct 2019

Ministry of Finance (2017b) Adherence of the Government Pension Fund Global (GPFG) to the Santiago principles. February draft. https://www.regjeringen.no/globalassets/upload/fin/statens-pensjonsfond/gapp/2017_gapp_adherence_gpfg.pdf. Accessed 5 Oct 2019

Moses JW (2020) Norway's Sovereign Wealth Fund. In: Pereira E, Spencer R, Moses J (eds) Experiences of managing wealth, CSR and local content policy: Sustainable development of extractive resources industries. Springer, New York (in press)

Moses JW, Letnes B (2017a) Managing resource abundance and wealth: The Norwegian experience. Oxford UP, Oxford

Moses JW, Letnes B (2017b) Breaking Brent: Norway's response to the recent oil price shock. Journal of World Energy Law and Business 10:103–116

NBIM (2015) Government Pension Fund Act. Unofficial translation from Norwegian. https://www.nbim.no/en/organisation/governance-model/government-pension-fund-act/. Accessed 5 Oct 2019

NBIM (2016a) Our voting records. https://www.nbim.no/en/responsibility/our-voting-records/. Accessed 5 Oct 2019

NBIM (2016b) Retningslinjer for observasjon og utelukkelse fra Statens pensjonsfond utland. https://www.nbim.no/no/organiseringen/styringsmodellen/retningslinjer-for-observasjon-og-utelukkelse-fra-statens-pensjonsfond-utland/. Accessed 11 Nov 2019

NBIM (2017) A trillion dollar fund. https://www.nbim.no/en/the-fund/news-list/2017/a-trillion-dollar-fund/?_t_id=1B2M2Y8AsgTpgAmY7PhCfg%3d%3d&_t_q=trillion+dollar&_t_tags=language%3aen%2csiteid%3ace059ee7-d71a-4942-9cdc-db39a172f561&_t_ip=66.165.1.186&_t_hit.id=Nbim_Public_Models_Pages_NewsItemPage/_7ea5e620-5eae-4ec3-a8d8-5ad4de19973e_en-GB&_t_hit.pos=1. Accessed 7 June 2019

NBIM (2018) Annual report. Government Pension Fund Global. Nr. 21. NBIM, Oslo
NBIM (2019a) Observation and exclusion of companies. https://www.nbim.no/en/responsibility/exclusion-of-companies/. Accessed 5 Oct 2019
NBIM (2019b) Decisions to exclude companies from the Government Pension Fund Global. https://www.nbim.no/en/the-fund/news-list/2019/decisions-to-exclude-companies-from-the-government-pension-fund-global/. Accessed 5 Oct 2019
NBIM (2019c) Strongest quarter ever. https://www.nbim.no/. Accessed 7 June 2019
NBIM (2019d) About us. https://www.nbim.no/en/organisation/about-us/. Accessed 11 Nov 2019
Norges Bank (2019) The executive board. https://www.norges-bank.no/en/topics/about/Organisation/The-Executive-Board/. Accessed 11 Nov 2019
Norwegian Government (2018a) Market value: Government Pension Fund Global (GPFG). https://www.regjeringen.no/en/topics/the-economy/the-government-pension-fund/government-pension-fund-global-gpfg/market-value-and-capital-inflow/id696852/. Accessed 5 Oct 2019
Norwegian Government (2018b) Market value: Government Pension Fund Norway (GPFN). https://www.regjeringen.no/en/topics/the-economy/the-government-pension-fund/government-pension-fund-norway-gpfn/market-value-gpfn/id710461/. Accessed 5 Oct 2019
NOU (1983). Petroleumsvirksomhetens framtid: det framtidige omfanget av petroleumsvirksomheten på norsk sokkel. Norges Offentlige Utredninger, Oslo
NOU (2017). New Central Bank Act. Norges Offentlige Utredninger, Oslo
NPD [Norwegian Petroleum Directorate] (2018) The market value to the Government Pension Fund Global, 1996–2017. Norwegian Petroleum. https://www.norskpetroleum.no/en/economy/management-of-revenues/. Accessed 5 Oct 2019
Qvigstad JF (2012) Om å Forvalte Rikdom. Speech to Det Norske Videnskaps-akademi. Nytt Norsk Tidsskrift 1:78–87
Skredderberget A (2015) Usannsynlig rik. Historien om Norge og Oljefondet. Kagge Forlag, Oslo
St. meld. nr. 29 (2000–2001) Retningslinjer for den økonomiske politikken. Finansdepartementet, Oslo
St. meld. nr. 25 (1973–74) Petroleumsvirksomhetens plass i det norske samfunn. Finansdepartementet, Oslo
SWFI (2018) Sovereign Wealth Fund rankings. November 2018 update. https://www.swfinstitute.org/sovereign-wealth-fund-rankings/. Accessed 24 Nov 2018

Old Fund, New Mandate: Saudi Arabia's Public Investment Fund (PIF)

Sara Bazoobandi

Introduction

Natural Resource Funds (NRFs) or resource-based Sovereign Wealth Funds (SWFs) serve a range of purposes for their sovereign owners, such as intergenerational saving, budget stabilization in times of revenue volatility, return maximization, absorption of excess liquidity, social development, economic diversification, and long-term political strategy. Among the oil-rich Arab countries in the Persian Gulf, preserving the wealth generated from the export of commodities for future generations who may not be able to generate the same level of wealth from commodity exports, has been one of the key purposes of SWFs. Resource-rich Arab countries seek to protect against future scenarios in which export revenue decline could occur, if resources run out, production decline occurs due to increasing domestic consumption, or technology develops in such a way that hydrocarbon resources lose significant export value. For those countries, such an intergenerational saving element is also meant to preserve the stability of the political system. The priority in these countries therefore is to use the resources to shape, reform, and/or advance economic, social, and political agenda of the next generation of

S. Bazoobandi (✉)
German Institute for Global and Area Studies, Hamburg, Germany

E. Okpanachi and R. Chowdhari Tremblay (eds.), *The Political Economy of Natural Resource Funds*, International Political Economy Series,
https://doi.org/10.1007/978-3-030-78251-1_9

the region's leaders. For example, in the UAE, Crown Prince Mohammed bin Zayed (Kirkpatrick 2019), has been spending sovereign wealth on regional issues such as the war in Yemen (Cornwell and Carvalho 2019), support of various political groups in Egypt (Emirates News Agency 2019a) and Syria (Emirates News Agency 2019b) to portray himself as an influential strategic leader. This trend has also been evident in the oil-rich Arab countries of North Africa. For example, in Libya, Seif Ghathafi was pursuing the same strategy when he took over the strategic planning of Libya Investment Authority (BBC News 2017) and the expansion of its diversified global portfolio.

In Saudi Arabia, Crown Prince Mohammad bin Salman has initiated major reforms to the kingdom's sovereign wealth management strategy, which are widely understood to be parts of his succession plan (Ghafar 2018). Saudi Arabia's Vision 2030 outlines the crown prince's vision for the future of the Saudi economy and reveals his desire to monopolize control of Saudi SWF, the Public Investment Fund (PIF), in order to support his own domestic and international agenda. Within the framework of Vision 2030, Saudi Arabia, under the leadership and vision of the crown prince, is planning to develop domestic projects funded by the PIF's assets in order to reduce the country's reliance on oil (Kingdom of Saudi Arabia 2017). The crown prince's proposed reforms are meant to transform one of the most conservative autocracies in the world into a force for global and regional development. These planned changes are unique in Saudi Arabia's history in that they aim to bring about unprecedented structural shifts in the kingdom's economy and sociopolitical landscape. The crown prince, with the assistance of his close advisors and various multinational consultancy firms and investment banks, is considered the single most important driving force behind these reforms (Azhar and Kalin 2019). The success of these reforms heavily depends on a number of factors, including, most significantly, foreign economic engagement with Saudi Arabia and political stability of the kingdom.

This chapter reviews the history of sovereign wealth and Saudi institutions that have been responsible for managing that wealth since the discovery of oil. It then explores the development of the PIF, the role of the PIF in Vision 2030, and the broader diversification policies in Saudi Arabia.

Oil and Social Contract in Saudi Arabia

The creation of Saudi Arabia in the twentieth century coincided with the discovery of oil, which has significantly influenced the evolution of Saudi Arabia's social, economic, and political structures. In 1938, six years after Abdulaziz Ibn Saud established the Kingdom of Saudi Arabia, oil was discovered, against the backdrop of a scramble for oil across the Middle East region in the interwar years. A group of US geologists, accompanied by Saudi officials, made the original find in the kingdom's Eastern Province. By 1941, oil fields were developed more formally, under the US-controlled company Arabian American Oil Company (ARAMCO). Over the years, Ibn Saud negotiated various profit-sharing agreements and bought stakes in ARAMCO. As oil revenues poured into the royal purse, large-scale infrastructure projects were undertaken and social spending, such as setting up an education system, led to fast-paced development and economic growth in the kingdom. In 1988, ARAMCO was completely bought out by the Saudi state and rebranded as Saudi Aramco, becoming a leading oil company and playing a crucial strategic role in the country's development to this day (Al Shurafa 2019).

Following the discovery of oil, Saudi rulers introduced a new social contract that was reinforced by a series of treaties with the British Empire that were introduced across the region (Onley 2007). These treaties allowed the newly established Arab states of the Persian Gulf to gain international recognition, and also enabled the ruling elite to gain the support of the local tribes (Onley and Khalaf 2006). As a result, the Saudi rulers successfully created a stable, peaceful sociopolitical environment in which various tribes were united and the ruling families became the sole providers of security, justice, and economic support. In return, the local tribal communities pledged their loyalty to the House of Saud.

Today, Saudi Arabia is one of the world's largest oil exporters, and its oil reserves are among the largest in the world. It still relies predominantly on these resources to drive economic development in the kingdom and fund the lifestyles of its relatively large population. Indeed, a complex system of wealth distribution and job creation programs serves, as in other oil exporters in the region, to underpin a longstanding social contract between the rulers and Saudi nationals. This system operates on a quid pro quo basis, in which citizens submit to political acquiescence and social conformity in return for public financial support. National resources are used to fund not only current budgetary needs but also asset accumulation strategies that seek to preserve this social contract for future generations.

The political structure of the kingdom has remained unchanged since it was founded. Saudi Arabia and other members of the Gulf Cooperation Council (GCC) are among the last remaining absolute monarchies in the world. Political power in Saudi Arabia has been monopolized by the members of the House of Saud, as seen in Table 1, and the division of political power has remained unchanged for many decades, as has Saudi Arabia's social contract. The king, his crown prince, and a group of his close advisors have traditionally been the key policymakers of the country. In this system, the political participation of Saudi citizens has been extremely limited, and access to political power has been dependent on personal links or family ties with the House of Saud.

Political power has been retained by the sons of the late King Abdulaziz Al-Saud, and there are no institutions outside of the Saudi Royal Court that holds legislative power. In the early 1990s, in response to repeated demands from educated and liberal Saudis and some members of the royal family, the late King Fahd promised political reforms in the country. A key element of these reforms was the establishment of a consultative council to allow other voices to be reflected in the country's affairs. Although the concept of consultation between the Saudi kings and tribal elites has existed since the establishment of the kingdom, the Consultative Assembly of Saudi Arabia (the Shoura Council, in its modern form) was launched in 1993. The Council is an advisory body, responsible for providing consultation to the king. All 150 members of the Council are directly appointed by the king for four-year terms. In 2013, women

Table 1 Saudi sovereign wealth fund deals (2012–2018)

Year	*Number of deals*	*Total value ($billion)*
2012	20	1.12
2013	7	0.448
2014	26	0.954
2015	10	3.05
2016	11	4.02
2017	9	73.2
2018	12	7.91

Source Sovereign Wealth Fund Institute, as quoted in Algethami (2018)

were included in the Council for the first time, with the appointment of thirty women including two Saudi princesses (BBC 2013). However, the council does not have any executive power, and its primary task is to study and propose laws.

The Council of Ministers acts as the official executive branch of the government, and the king as the Prime Minister. All Saudi ministers are appointed and dismissed by royal decrees issued by the king. In practice, the Royal Court holds ultimate control over legislative matters. Influential members of the royal family, ministers, some of the king's special advisors, influential religious scholars, and tribal leaders can all be instrumental in forming royal decrees. Article 24 of the Law of the Council of Ministers delegates full executive power to the council over various issues such as creating and organizing public institutions and following up on the implementation of the general development plans (fanack.com 2020). All laws proposed by the Shura Council must be passed by both the Council of Ministers and the Prime Minister (i.e., the King), before they can be introduced as official laws. In case of disagreement between the Council of Ministers and the King, the King will have the final decision-making power.

During King Abdullah's reign, municipal councils were created as part of the wider political reform in Saudi Arabia. The local council members are elected through a direct voting system. Local elections were held in 2005 and 2011, and in 2015 women were allowed to participate. The election turnouts have been significantly low: in the 2015 election, only a quarter of the Saudi population participated. The municipal councils have limited power that does not go beyond advising local governments and helping oversee budgets. However, the last local council election was considered a significant political event as it allowed women to become involved in politics. This decision received strong opposition from the country's top religious leaders, who saw it as "opening the door to evil" (Lillywhite 2015).

Saudi Arabia's Sovereign Wealth from a Historical Perspective

Institutional investors such as Kuwait Investment Authority, Abu Dhabi Investment Authority, and Qatar Investment Authority have been operating in the region for many decades, and their organizational structures, size, global reputation, and investment strategies have evolved over time.

Some of the GCC countries have introduced specific rules and mechanisms to regulate and implement savings, investment, and withdrawal strategies for SWFs' assets. In Saudi Arabia, however, sovereign wealth management has remained considerably underdeveloped in comparison to other GCC economies. In fact, until recently, Saudi Arabia, unlike other oil-rich economies of the GCC, did not have an official government investment vehicle that fitted the definition of SWFs. The Saudi Arabian Monetary Agency (SAMA), the country's central bank, has been considered an SWF in various studies; SAMA's foreign investments, mainly in US Treasury bonds, were the Saudi sovereign investment vehicles that came closest to the definition of SWFs. Indeed, for decades, SAMA was seen as a double-purpose institution that served both as Saudi Arabia's central bank and a long-term sovereign wealth institutional investor. Yet, SAMA has been one of the most conservative sovereign investment vehicles in the Arab Gulf nations. In contrast to other regional sovereign wealth funds such as Abu Dhabi Investment Authority and Kuwait Investment Authority, which have maintained highly diversified portfolios with some high-risk adjusted-return assets, SAMA foreign exchange reserves were mainly held in dollar-dominated fixed-income securities (Bazoobandi 2013). The PIF was also not considered an SWF as almost all of its investments were inside the kingdom, whereas SWFs' assets are held mainly outside of their sponsoring governments. Thus, while Saudi Arabia has experienced various periods of oil-price-led economic booms and has, therefore, succeeded in accumulating financial buffers from surplus commodity export revenues, its overall oil wealth management has remained behind its regional peers in terms of the size of assets under management of the SWFs, the diversity of assets portfolios, and the institutional development of the funds.

A key historical dynamic that has played a significant role in shaping Saudi Arabia's oil management and the evolution of its sovereign wealth fund is Saudi-US relations. These relations were forged in the development of the original oil concessions, and a strong relationship at the political level developed alongside these burgeoning commercial ties. In the early 1940s, Saudi Arabia was faced with growing British hegemony in the region and related preoccupations about potential British attempts to gain control of oil supplies. Ibn Saud, therefore, came to an agreement with US President Roosevelt both to counterbalance British power and to secure valuable protection. In return for access to Saudi Arabia's oil resources, the USA offered security guarantees to the kingdom. US military airfields were constructed in Saudi Arabia, and US protection was given to both oil fields and the pilgrimage routes to Mecca and Medina. The US-Saudi alliance was further bolstered by a mutual drive to counter the rising communist threat and the Shia Islamic Republic of Iran. These examples demonstrate that throughout their relationship, the US and Saudi Arabia have found common ground through aligned security perceptions and development needs, rather than a set of shared values. The relationship has proved durable, in spite of difficult moments such as US support for Israel in the Yom Kippur War or the Saudi involvement in the 9/11 terrorist attacks in New York.

In recent years, however, uncertainty over the viability of the alliance owing to changing global conditions has risen. On one side, the importance of Saudi oil supplies to the USA has declined in the wake of rising domestic energy development in the USA. Recent developments in the US energy sector have changed it from a net importer of energy to a major exporter. Although Saudi Arabia maintains primacy in terms of oil exports and its reserve position on the global stage, the growth of both alternative energy sources and non-OPEC oil and gas suppliers has raised questions about sustainability in the longer term and concerns over the impact of lower oil prices on the kingdom's fiscal position.

On the other side, there are signs that Saudi Arabia may increasingly pivot toward more like-minded allies in the future. US leaders have made great efforts to prevent controversies such as the murder of *Washington Post* journalist Jamal Khashoggi in 2018 from affecting the US-Saudi relationship. However, public and congressional outcry in the US against such controversies and the general human rights records of the kingdom, along with the rising power of regimes in China and Russia whose values and styles of rule align more with Saudi Arabia's outlook, could increasingly draw Saudi Arabia outside the US sphere of influence and offer the kingdom alternative security guarantors. In the case of China, growing energy needs also offer a more promising substitute export market for Saudi oil.

Signs of such tensions in US-Saudi relations were seen in 2016, when Saudi Arabia threatened to dump US assets worth hundreds of billions of dollars if Congress passed a proposed bill that would pave the way for the Saudi government to be held responsible in the US judicial system if found to have participated in any way in the 9/11 terrorist attacks (Mazzetti 2016). This episode exposed the US-Saudi alliance's critical codependence on Saudi holdings of US debt. This dynamic emerged in the 1970s after the collapse of the gold standard and the establishment of the petrodollar system. The main condition of the US military protection that Saudi Arabia was offered in the 1973 US-Saudi agreement was a commitment to pricing oil in US dollars. This created large US dollar reserves for Saudi Arabia, which funded the country's fiscal needs and were used to invest in "safe haven" US dollar-denominated assets and US Treasury bills (USTs). This, in turn, funded US government spending by buying its government debt and made the US dollar the world's reserve currency.

Uncertainty around a potential future shift in this well-established paradigm has grown in recent years. The risk of persistently lower oil prices could force Saudi Arabia to draw down its UST holdings to fund domestic spending needs. In addition, various geopolitical factors could potentially lead to a sell-off of Saudi-held USTs and other US-based or US-dollar-denominated assets. The aforementioned risk of retaliatory asset dumping if the US is seen to be overly critical of the Saudi regime also remains a real threat. In addition, the growth of non-OPEC producers and the existence of sanctions that prevent some producers from selling their oil in dollars has already led to instances in which oil is traded in other currencies, such as Chinese Yuan (Cho and Kumon 2019). A more wholesale shift into trading in alternative currencies could lead to a shift in perceptions toward the US dollar as the global reserve currency, as well as to the rise of "challenger" currencies from other global economic powers. However, the association of most Middle Eastern oil producers' currency with the US dollar, the alignment of their monetary policy stances with the US Fed, and the primacy of the US dollar in the global financial system are strong counterbalances, at least for the moment, to any meaningful shift away from the petrodollar and the reserve status of the US dollar.

Perceptions of Middle Eastern oil exporters in global markets changed significantly during the financial crisis of 2008. As global assets slumped, the foreign portfolios of Arab oil exporters fell in value, but these investors were also offered opportunities, as low asset prices and a dearth of liquidity and capital in global markets increased their financial clout. As the financial crisis unfolded, the ready liquidity of several Middle Eastern SWFs was crucial in stemming the crisis that was ripping through the global banking sector. A series of capital injections into ailing but systemically significant financial institutions and key infrastructure, such as Qatar Investment Authority's stake in Barclay's Bank, raised the profile of these funds. During this period, the investment strategies of some SWFs were shown to be far from the low-risk, long-term approaches typically associated with this class of investors. These more proactive approaches, seen from the likes of the Qatar Investment Authority and the Dubai funds, Mubadala and Dubai Investment Corp., illustrated how the spoils of the earlier oil boom were being used to drive much bolder and more proactive investment strategies than had been assumed previously. Although they were financially necessary at the time, these entities soon faced political backlash. Reservations about the objectives of SWFs had been apparent prior to the crisis, most prominently in the Dubai Ports World controversy in 2006. However, the increased activity of SWFs during the crisis drew more attention to their lack of transparency and poorly defined mandates, giving rise to heightened national security concerns.

This backlash also coincided with other market and socioeconomic factors that caused SWFs to step back from further risky investments in international markets. Falling oil prices and the impact of the global downturn on regional growth—and, later on, the destabilizing effects of the Arab Spring uprisings—necessitated higher levels of domestic spending while also creating tighter fiscal conditions.

In 2009, against the backdrop of post-crisis backlash against Middle Eastern sovereign wealth funds, and with depressed asset valuations to be found in global markets, Saudi Arabia officially set up a new $5.3 billion SWF, Sanabil el-Saudi. Its focus was announced to be primarily related to the country's existing areas of expertise, especially plastics (Merzaban 2009). This was the kingdom's first formal SWF and a late arrival to the Arab SWF landscape. Previously, management of Saudi Arabia's oil wealth had been undertaken by the investment arms of SAMA and its Finance Ministry, SAMA Foreign Holdings. Following the announcement of Sanabil al-Saudi's launch, little more was heard about the fund's strategy or individual investments (Monk 2010). In fact, until the rebranding of the Public Investment Fund (PIF) as the country's primary sovereign wealth fund in Crown Prince Mohammed bin Salman's Vision 2030, it appears likely that management of the kingdom's resource wealth still remained largely with SAMA.

2030 Economic Vision, Rebranding the Public Investment Fund as the Saudi SWF

Saudi Arabia's newly introduced 2030 Vision has put a strong emphasis on the importance of the Public Investment Fund (PIF) in achieving the country's economic goals. The PIF grew in prominence when it was reintroduced as the official sovereign wealth fund in Vision 2030. The official document (Kingdom of Saudi Arabia 2017) that outlines the government's plans for development of the PIF starts with two quotations from the King:

> We shall be working towards building a solid economy based on strong grounds that will lead to the multiplication of the sources of income.

and his crown prince:

> Our nation holds strong investment capabilities, which we will harness to stimulate our economy and diversify our revenues.

These quotations are clear indicators of the PIF's strategy in the years to come. Saudi leadership is seeking one key economic objective through the PIF's investments: to diversify the government income away from oil. The Saudi economy, particularly government revenue, has been historically dependent on oil export income. As a result, both the country's GDP and the government budget deficit/surplus have been directly linked with oil price fluctuation.

The government has officially outlined four main objectives for the PIF in accordance with Vision 2030 (Kingdom of Saudi Arabia 2017):

1. Grow the assets of the PIF;
2. Unlock new sectors through the fund;
3. Build strategic economic partnerships through the fund;
4. Localize cutting-edge technology and knowledge through the Public Investment Fund.

The new government's strategy has also defined a number of Key Performance Indicators (KPIs), in order to measure the performance and impact of the PIF activities. The government is planning to (Kingdom of Saudi Arabia 2017):

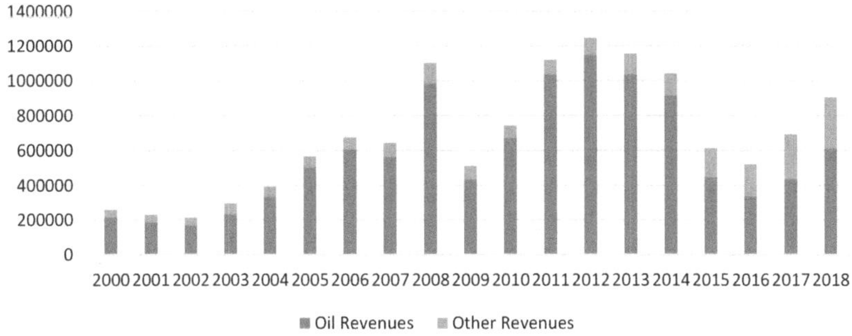

Fig. 1 Share of oil revenue from total revenue (million riyals) (*Source* Saudi Arabia Monetary Agency Annual Data [2018])

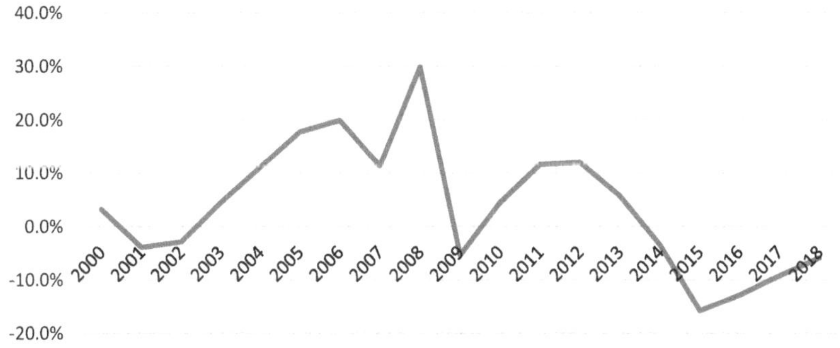

Fig. 2 The ratio of deficit/surplus to GDP (*Source* Saudi Arabia Monetary Agency Annual Data [2018])

- Increase the PIF's assets under the management to SAR 1.5 trillion in 2020 (around $0.4 trillion) from a baseline of SAR 0.84 trillion in 2017.
- Invest 20% of the PIF assets in new sectors by 2020. The fund is also aiming to generate SAR 30 billion direct contribution to GDP from the new sectors.
- Invest 25% of the fund in international assets (from a baseline of 5% in 2017).
- Create 11,000 direct high-skilled jobs in cutting-edge technology and knowledge sectors. Between 2018 and 2010, the fund is also

planned to invest SAR 210 billion in technology research and development.

- The fund is aiming to generate up to 6.3% of the country's total GDP by 2020, up from 4.4% in 2016.
- In total, the fund's investment projects are aiming to create 20,000 direct jobs, 9,000 indirect jobs, and 256,000 jobs in the construction sector. The fund is also seeking to boost household consumption rate by 1.4 pp by 2020. The fund's financial activities are also expected to increase the inflation rate by about 0.7% during the same period.

Relaunching the Public Investment Fund

The PIF was established in 1971, aiming to support the state's strategic business interests. Unlike SAMA, which was in charge of the bulk of the Saudi government's foreign investments, PIF was mostly focused on the domestic market. The PIF holds about $150 billion of its assets in listed Saudi companies. It has been behind the creation of some of the country's most successful economic champions, including the Saudi Basic Industries Corporation (Sabic), Saudi Telecom Co., and the largest lender in Saudi Arabia, National Commercial Bank (Algethami 2018). In 2015, the year that Vision 2030 was launched, the government announced major steps in overall management and future strategy of the PIF. The fund's supervision was delegated to a newly established government body, the Council of Economic and Development Affairs (CEDA). Prior to this, the fund was managed by the Ministry of Finance. CEDA was formed by the Saudi Council of Ministers and is responsible for implementing necessary measures to facilitate achieving the Vision goals. CEDA has developed "a comprehensive governance model aimed at institutionalizing, enhancing its work, facilitating the coordination of efforts among relevant stakeholders and effectively following-up progress" (Kingdom of Saudi Arabia 2017). The Council consists of the Crown Prince, Mohammad bin Salman, who is the President of CEDA, and 20 Saudi ministers. There are three other council members: the chairmen of the Board of Directors for Local Content and Government Procurement Commission and the Board of Directors of the Saudi Commission on Tourism and National Heritage, and the Head of Experts Commission at the Council of Ministers. It is widely accepted that the crown prince holds a significant power at CEDA, and most of the Council's decisions are influenced by the crown prince and his closest allies.

Through an aggressive growth and investment strategy, the PIF is currently involved in a number of new investment projects in sectors such as military, industry, e-commerce, waste management, entertainment, real estate, and infrastructure. The government is planning to attract $5.3 billion in foreign direct investments by 2020 through the PIF. As a result of engaging foreign investors, the PIF's source of funding is expected to be different than that of traditional Arab SWFs that receive their assets by direct government allocations from excess commodity export revenue. Instead, the government is planning to raise funds through a combination of capital injections from the source of oil surplus revenue, assets transferred from other government investment institutions, loan and debt instruments, and investment returns (Chandrasekaran 2017). The fund is reportedly planning to sell 70% of its shares in Sabic to Aramco in order raise $70 billion. In 2018, the PIF raised $11 billion through borrowing (Algethami 2018). In July 2019, the *Financial Times* reported that PIF agreed to the initial conditions of another $10 billion short-term loan from a group of 10 international banks including Bank of America Merrill Lynch, BNP Paribas, and Citi Group. The loan was structured to be repaid by the end of 2019 after the first Initial Public Offering (IPO) of Aramco. The government plans to increase the assets under the management of PIF from $300 billion to $400 billion by the end of 2020 to $2 trillion by 2030 (Kerr 2019).

In recent years, the PIF has been entering new markets and has made several high-profile investments, including the following:

- In electric car manufacturing companies: $2 billion Tesla and $1 billion in Lucid Motors
- In e-commerce: $3.5 billion investment in Uber
- In infrastructure: $20 billion in Black Stone Group
- In technology: $45 billion in Soft Bank Group (Kerr 2019; Raval and Massoudi 2019)

As noted, the PIF is different from other Saudi sovereign investment institutions such as SAMA, and also from other Arab SWFs in many respects. The source of funds, investment strategy, and financial and strategic goals of the PIF are distinctive and unique in that no other institutional investor in the region has combined such levels of financial clout and strategic ambition in one organization. The PIF is not specifically aimed at intergenerational savings as are other global SWFs such as the Norwegian SWF (Gross 2017). Instead, the fund is meant to act as a Stabilization Fund, similarly to Russian SWFs, in order to provide a protection mechanism for the Saudi economy against the fluctuation of oil prices. Moreover, the fund is tasked with helping to promote competitiveness in the Saudi economy by a combination of direct investment and attracting foreign companies. The latter is aiming to support the transfer of technology and skills into the Saudi economy (Di Paola 2019). Recent IMF reports have welcomed some of the government's economic strategies that are promoted through the PIF's activities. As such, the IMF acknowledged the measures that support entering new sectors through foreign partnerships, such as by entering into joint ventures with leaders in the field, as the most expedient route to acquire the technological and methodological know-how needed to enter into new sectors and a useful tool for the economic diversification strategy of Saudi Arabia (IMF 2019). The PIF is actively seeking to adopt such approaches to foster economic development and diversification through strategic partnerships. For example, tenders for a dozen renewables energy projects were announced in early 2019 (El Gamal and Carvalho 2019), on which UAE renewables giant Masdar is expected to take partnership (Rahman 2019).

The PIF is meant to serve as a special purpose investment vehicle that will align with and complement other components that are required to promote Crown Prince Mohammad bin Salman's domestic and international agenda. The crown prince has invested heavily in improving the country's image (Waterson 2018), seeking to present himself as a young open-minded leader with a vision, and the capacity, of turning Saudi Arabia into a global economic power. Since the launch of Vision 2030, which assigned this set of strategies to the PIF, all decisions related to the future of the fund have been directly controlled by the crown prince. The *Financial Times* has quoted an anonymous source who claimed, "The PIF is the vehicle for everything," a significant amount of government assets are planned to go through the PIF, and "it will be moulded and shaped in any way that is required to help deliver his highness's orders" (Raval et al. 2019).

The crown prince has appointed a number of his advisors to strategic management positions at the PIF. In September 2015, he appointed Yasir Al-Rumayyan, one of his most trusted advisors, as the managing director of the PIF. Before entering the Royal Court, Al-Rumayyan served in senior management positions at the Saudi Stock Exchange, the Capital Markets Authority, and Saudi Fransi Capital. He survived several cabinet reshuffles and changes in the crown prince's economic team, becoming an advisor to the Cabinet and a member of the board of the Saudi Industrial Development Fund (SIDF) in 2016 (Arab News 2019). Considering his senior banking and finance background, Al-Rumayyan seems to hold the required expertise for this position, in addition to the crown prince's trust. He appointed a team of banking managers from various global banks to handle the day-to-day operations at the PIF. Given the timeline of Saudi Arabia's economic vision, the PIF's management team has been tasked to implement necessary changes in a relatively short span. As a result, the number of employees increased from 40 to more than 700 over the course of three years, with plan for this number to increase to more than 1000 employees by the end of 2020 in line with the PIF's aim to increase its investment to $400 billion this year (Martin 2020).

As noted, the Saudi government is planning to increase the size of assets under the management of the PIF from a combination of sources, including proceedings from privatizing government-owned companies. Unsurprisingly, privatization plans are closely linked with the future development and strategic planning of the fund. The government has announced various measures for privatizing state-owned companies, with more than 20 companies in various sectors, such as agriculture, energy, and sports, scheduled to be privatized in 2019. Selling government-owned companies is expected to generate around $10 billion and create up to 12,000 jobs by 2020 (Rashad 2018).

The centerpiece of privatization and the government's effort to attract foreign direct investment was the Initial Public Offering (IPO) of a 5% stake in Saudi Arabia's national oil company, Aramco. The process has been postponed on a few occasions by the government since its initial announcement in January 2016. In an interview in June 2019, the crown prince announced the plan as the world's biggest IPO (Gross 2017) and confirmed that, after some delays, it would finally take place in 2020–2021 (Di Paola 2019). Although the new Saudi economic agenda seeks to empower the private sector by promoting privatization and attracting foreign direct investments, the complexities of the Saudi market, deriving from its political structure, judicial system, and market dynamics, have not been seen as particularly "inviting" to foreign investors (Al-Sulayman, 2018). Events such as the diplomatic crisis with

Qatar, the war in Yemen, the ongoing crisis with Iran, and, for a brief period, the murder of Jamal Khashoggi, have had negative effects on the global perception/confidence of investments in Saudi Arabia, as the gap widens between the expectations of the Saudi government and the interests that motivate international investors. The government has been reportedly seeking $2 trillion in value for Aramco's IPO, but international analysts suggest that the total value for the company cannot surpass $1.2 trillion. Moreover, the overall economic prospect of investments in Saudi Arabia has remained unattractive, since IMF forecasts show that the 2019 growth is not expected to rise above 0.2% (Barbuscia 2019). Therefore, the overall level of FDI inflows to Saudi Arabia does not look promising (Mogielnicki 2019).

The PIF's expansion plans and related reforms are premised on attracting foreign investors, something that is very much dependent on how events unfold at regional and domestic levels for Saudi Arabia and whether sufficient investor appetites can be attained. The Aramco IPO has great symbolic importance for the kingdom's broader development strategy and also seeks to address imminent funding needs at the center of the Vision 2030 strategy. The government hopes the 5% stake will raise $100 billion. Funds generated from the initial stage of Aramco's privatization are necessary to fund the PIF's investment strategy, as its current commitments cannot be serviced by its existing illiquid assets, and the government faces serious fiscal constraints due to lower oil prices. Public expenditures required for the high public sector wage bill and costly social spending commitments do not leave much space for additional fiscal contributions to PIF projects. Investments from the government budget would negatively affect the kingdom's already precarious fiscal position and investment perceptions.

Saudi Arabia's SWF: An Assessment

The overall management of Saudi sovereign wealth has been relatively transparent. SAMA has been one of the best structured organizations in Saudi Arabia, and the PIF's assets have been protected from corruption issues. Nevertheless, corruption has existed across various areas of the Saudi economy, and tackling it has been a great challenge for the government. The crown prince launched a major anti-corruption incentive in 2017, which aimed to address deep-rooted issues that had reportedly allowed up to 10% of all Saudi government spending to evaporate fraudulently every year (BBC 2019).

One of the most controversial steps in Saudi Arabia's anti-corruption campaign was the arrest and imprisonment of prominent business and political figures at the Ritz Carlton Hotel in Riyadh in 2017. A former official described the Ritz events as a move "to remind people going forward that their wealth and their well-being would depend on the crown prince and not anything else" (De Luce et al. 2018). The Saudi Royal Court announced that a committee led by Crown Prince Mohammed bin Salman had recovered $107 billion in settlements, with total of 381 individuals summoned by the commission (Seznec 2020). Some of the wealthiest, and previously most influential, Saudi figures were among those involved in the settlement cases. Bin Ladin Group, a well-known construction company with close ties to the Royal Court, had transferred some of the holdings' shares to the state. This was after the company's chairman, Bakr Bin Ladin, and several Bin Ladin family members were detained in the crackdown. Prince Miteb bin Abdullah, once seen as a leading contender for the throne, was freed after paying more than $1 billion. Prince Alwaleed bin Talal, chairman and owner of global investment firm Kingdom Holding, was also arrested but the details of his settlement were not made public (Rashad et al. 2019).

The crown prince has also appeared keen to improve perceptions of the transparency of key Saudi institutions. Notably, prior to the Aramco IPO, the government began to take steps to clarify the company's financial links with the energy ministry. This may well have been to protect the company from the potential legal risks of financial misconduct after the IPO. One major issue that was raised during this process was the tradition of Aramco paying the expenses of the energy minister. Former energy minister Khalid Al-Falih, who until 2019 was also Aramco's chairman, reportedly came under public scrutiny when his colossal personal expenses were reviewed as part of the broader initiative to improve the company's governance. For years, Aramco, the world's most profitable oil company, paid Al-Falih's expenses, including his travel costs, ultra-luxury hotels and using the company's private jet. Unsurprisingly, Al-Falih had to step down as Al-Rumayyan, who previously was the head of the PIF, took over Aramco (IMF 2019). Ahead of Aramco's much-anticipated IPO, the government announced that the company is officially separated from the ministry of energy. This decision was also used as a justification to remove Al-Falih from Aramco without any official commentary about the personal expenses scandal.

As noted above, the Saudi institutions that have been in charge of managing most of the country's sovereign wealth have maintained fairly transparent operations. There are, however, no clear regulations to clarify the government's commitment to saving strategies for Saudi Arabia's oil wealth. Under the leadership of the crown prince, the Saudi government has been seeking speedy policies to expand the size of the country's SWF. This indicates that the recent government efforts may well be aimed at compensating for the years of high government expenditure and low sovereign wealth savings. The government has announced plans to merge Sabic and Aramco, facilitated by Aramco's purchase of 70% of Sabic from the PIF (Roll 2019). Plans of this kind confirm that the authorities are seeking quick alternative strategies to inject much-needed liquid assets into the PIF in order to meet the country's SWF targets. The merger, meant to boost Aramco's downstream growth prospects ahead of the IPO, is expected to raise $61.9 billion for the PIF (Seznec 2020). Since the launch of Vision 2030, the Saudi leadership has implemented a mixture of corporate strategy and government policymaking. The mixture of management consultant-led corporate strategy initiatives and civil servant-led public policy measures have created a sense of confusion, and is therefore unlikely to be able to deliver realistic and sustainable outcomes in the long run. As a result, the transformations in the PIF indicate that the new structure of the fund is not focusing on institutional capacity building. Instead, the changes are shaped and implemented as a result of direct orders from the crown prince and his team of financial advisors, which may create a hospitable environment for the decline of transparency.

As noted, the day-to-day affairs of the fund have been in control of the PIF's managing director, Yaser Al-Rumayyan. The senior governing body of the fund is the PIF's board of directors. The board consists of nine (appointed) members that include the PIF's managing director, the chairman of the board, Prince Mohammad bin Salman, and six Saudi government ministers of foreign affairs, commerce and investment, states, energy, finance, and economy and planning. The chairman of the board of directors for the Saudi Commission for Tourism and National Heritage is also on the board of the PIF. The central bank, SAMA, does not have any representation in the PIF governing body. Given the strong influence of the crown prince in forming the new structure of the PIF, this decision may indicate his plan to give the PIF a high degree of independence from the bureaucratic structure of the central bank. Although this arrangement may well be intended to increase efficiency, it can also potentially lead to corruption.

The fund does not publish quarterly figures on its website. Changes to the staff at the PIF can be opaque and subject to the decisions of certain powerful individuals. The investment strategy of the fund is not clarified in any of the official documents that have been released by the government through the fund's website or other government online platforms such as the Vision 2030 website. Although the government has placed great emphasis on the importance of Vision 2030 for the future economic transformation of Saudi Arabia and the role of the PIF in materializing the goals of the Vision, there is very little information on the PIF's management structure and investment mandate prior to 2015. Furthermore, the government has not clarified the reasons for choosing the PIF as the country's official SWF.

In developed democratic political systems, the ideological orientation of the government parties strongly influences the incumbent government's economic and financial decisions. Given that political power has been monopolized by the Saudi royals since the discovery of oil, sovereign wealth management and all other government economic decisions are controlled by the king and his trusted circle of economists and financial advisors. As noted above, political power is also tightly controlled by the Royal Court and the Council of Ministers, who are directly appointed by the king. This means that political fragmentation is dealt with at a very senior level. Therefore, disagreements over sovereign wealth management strategies are manifest neither at the vertical level between the municipalities and the national government nor at the horizontal level among ministers, central bank leaders, or other technical government teams. By the same token, citizen groups and extractive industry representatives are not included in the government's strategic decision-making concerning sovereign wealth management. A combination of the country's decades-long social contract and political power structure has created an environment in which the government is the sole owner of the wealth. As a result, certain individual royal family members, or their close acquaintances, have monopolized wealth management decision-making while the citizens benefit, or suffer, from the outcomes of their decisions. For many decades, the Saudi royal elite has played the role of the patron in a clientelist relationship in which Saudi citizens, in exchange for their political loyalty, receive the patron's direct and indirect financial supports. Although clientelist political relations are often associated with elections, the model very much applies to day-to-day state-citizen relations in Saudi Arabia. All in all, the political power structure in Saudi Arabia confirms that the overall sovereign wealth strategy is indeed controlled by the senior political elite. The two existing councils do not hold any significant power in shaping the country's sovereign wealth funds' savings plans, investment strategies, or withdrawal mechanisms. Moreover, Saudi citizens have no means of taking part in forming such policies: given the nature of its political system, the management of the country's sovereign wealth is not a topic that is raised in the electoral circles.

Conclusion

Government spending pledges made after 2011 to address social grievances in the wake of region-wide civil unrests, combined with the oil price decline of 2014, have magnified the Saudi government's existing financial challenges. These challenges have exerted immense pressure on Saudi policymakers to implement long-awaited economic and social reforms in an attempt to halt a declining fiscal position and put the kingdom on a more sustainable economic footing. As the number of Saudi youths in search of employment has increased and the country's wealth generated from hydrocarbon exports has declined, Saudi policymakers have felt an urgent need to implement policies to diversify the economy and pave the way for greater private sector growth in order to create jobs for the burgeoning young population. These changes are seen as crucial to releasing social pressures and protect the political establishment from potential social unrest similar to what took place in other Arab countries.

Structural reforms will be necessary, along with improvements to the kingdom's legal infrastructure, if perceptions of Saudi Arabia are to improve. Foreign investment is a necessary prerequisite to achieving the country's economic reform goals. Historically, the government has been the primary provider of jobs in Saudi Arabia. However, as the economic pressure of public sector wages on the government's budget has increased, it has become clear that the private sector's role must be strengthened in order to create jobs and generate economic growth. Executing the necessary changes to rebalance the labor market away from the public sector will be politically challenging. Wage levels will need to be reduced significantly in order for private-sector wages to become more attractive and Saudi labor more internationally competitive. Productivity is also considered low by international standards, meaning that the kingdom's education policy needs to continue to be refined to align more closely with the needs of international labor markets (IMF 2019).

Since 2016, the Saudi crown prince has been leading the economic policy through the ambitious Vision 2030 strategy. The government has promised economic diversification, measures to decrease the country's dependence on oil and the creation of an economic environment that allows the private sector to take a strong role and reform Saudi Arabia's bureaucratic system by privatizing public assets and fighting internal corruption. The private sector has, therefore, become the central element of the new policies that are branded as Vision 2030.

The expansion plans for the PIF lie at the center of Vision 2030. The Saudi economy is under mounting pressure caused by a combination of heavy reliance on oil income, high government expenditure, and fluctuating global oil markets. The government has planned to reintroduce the PIF by changing its mandate and substantially increasing the size of assets under its management. Saudi leadership, and specifically the crown prince, feels an urgent need to take quick measures to maintain the country's political and economic stability. The crown prince seems to be using these economic and social reforms not only to address existing challenges, but also to prepare for a smooth succession.

The Saudi government is trying to tackle various economic issues such as unemployment, a weak private sector, and heavy dependence on oil income by turning one of the country's largest state-owned funds, other than that of the central bank (SAMA), into the official Saudi SWF. As the central bank of Saudi Arabia, SAMA has been, by definition, in charge of the country's monetary and fiscal policies, and has therefore not developed sophisticated investment strategies, like those of the other GCC SWFs, aimed at maximizing investment returns. That said, the new structure of the PIF also does not fit those of classic SWFs: "Unlike, for example, the Norwegian SWF, which is managed by the Norwegian central bank on behalf of the Ministry of Finance, the PIF operates within the framework of an investment company model. It is both the owner and manager of the fund's assets" (Roll 2019).

The management structure and decision-making system at the PIF have become increasingly centralized since 2015. Considering the social and political structure of Saudi Arabia, Saudi citizens have no mechanisms for participating in decision-making processes about the Fund. This increases the risk of assets being used mainly to pursue Crown Prince Mohammed bin Salman's broader agenda to facilitate his succession. Finally, recent attacks on Saudi oil infrastructures have confirmed the country's broader future challenges. Therefore, rebranding the PIF as a fast-growing, aggressive global SWF, may not be the ultimate solution for addressing the long-term and short-term socioeconomic challenges faced by Saudi Arabia.

References

Abdel Ghafar A (2018) A new kingdom of Saud? Brookings Institutions. https://www.brookings.edu/research/a-new-kingdom-of-saud/

Algethami S (2018) What's next for Saudi Arabia's Sovereign Wealth Fund. Bloomberg. https://www.bloomberg.com/news/articles/2018-10-22/what-s-next-for-saudi-arabia-s-sovereign-wealth-fund-quicktake

Al Shurafa S (2019) All you need to know about Saudi Aramco. Gulf News, 18 Sept. http://gulfnews.com/world/gulf/saudi/all-you-need-to-know-about-saudi-aramco-1.1568799084771

Al-Sulayman F (2018) State business relations and the New Economic Agenda in Saudi Arabia. Rice University's Baker Institute for Public Policy, Issue Brief 07.18.18. https://www.bakerinstitute.org/media/files/files/96d3c6ac/bi-brief-071818-cme-carnegie-london3.pdf

Arab News (2019) Yasir Othman Al-Rumayyan, chairman of Saudi Aramco. https://www.arabnews.com/node/1549461/saudi-arabia

Azhar S, Kalin S (2019) Saudi Arabia's hometown ambitions could clip wealth fund's wings. Reuters, 28 June. https://www.reuters.com/article/us-saudi-pif-investment-insight/saudi-arabias-hometown-ambitions-could-clip-wealth-funds-wings-idUSKCN1TT0OE

Barbuscia D (2019) IMF slashes 2019 growth forecast for Saudi Arabia to 0.2% from 1.9%. https://www.reuters.com/article/us-saudi-economy-imf/imf-slashes-2019-growth-forecast-for-saudi-arabia-to-02-from-19-idUSKBN1WU22J

Bazoobandi S (2013) The political economy of the Gulf Sovereign Wealth Funds. Routledge, Abingdon and New York

BBC News (2013) Saudi Arabia's king appoints women to Shura Council. https://www.bbc.com/news/world-middle-east-20986428

BBC News (2017) Who is Libya's Saif al-Islam Gaddafi? https://www.bbc.com/news/world-africa-14617511

BBC News (2019) Saudi Arabia ends major anti-corruption campaign. 31 Jan. https://www.bbc.com/news/world-middle-east-47065285

Chandrasekaran A (2017) Saudi Arabia in investment spree mode. Gulf News, 29 Oct. https://gulfnews.com/business/markets/saudi-arabia-in-investment-spree-mode-1.2115094

Cho Y, Kumon T (2019) China, Russia and EU edge away from petrodollar. Nikkei Asian Review, 9 Jan. https://asia.nikkei.com/Economy/China-Russia-and-EU-edge-away-from-petrodollar

Cornwell A, Carvalho S (2019) UAE signs $5.5 billion in military contracts as Yemen war heightens scrutiny. Reuters, 21 Feb. https://www.reuters.com/article/us-mideast-defence-emirates/uae-signs-55-billion-in-military-contracts-as-yemen-war-heightens-scrutiny-idUSKCN1QA1CS

De Luce D, Dilanian K, Windrem R (2018) How a Saudi royal crushed his rivals in a "shakedown" at the Ritz-Carlton. NBC News. https://www.nbcnews.com/news/mideast/how-saudi-royal-crushed-his-rivals-shakedown-ritz-carlton-n930396

Di Paola A (2019) Saudi Crown Prince says Aramco's planned IPO is on track. Bloomberg. https://www.bloomberg.com/news/articles/2019-06-16/saudi-crown-prince-expects-aramco-ipo-in-2020-2021-paper-says

El Gamal R, Carvalho S (2019) Saudi Arabia sees domestic energy use falling, plans renewables push. Reuters, 15 Jan. https://www.reuters.com/article/us-saudi-energy-reforms/saudi-arabia-sees-domestic-energy-use-falling-plans-renewables-push-idUSKCN1P918N

Emirates News Agency (2019a) Shaikh Mohammad Bin Zayed in Egypt: Three MOUs signed. https://gulfnews.com/world/mena/shaikh-mohammad-bin-zayed-in-egypt-three-mous-signed-1.1553692969262

Emirates News Agency (2019b) UAE aid to Syria reaches Dh3.59 billion from 2012–2019. https://www.khaleejtimes.com/nation/abu-dhabi/uae-aid-to-syria-reaches-dh359-billion-from-2012-2019

fanack.com (2020, September 29) Politics of Saudi Arabia. https://fanack.com/saudi-arabia/politics-of-saudiarabia/politics-of-saudi-arabia/

Gross S (2017) The Saudi Aramco IPO is a game-changer for the Saudi economy. Brookings Institutions. https://www.brookings.edu/blog/markaz/2017/06/06/the-saudi-aramco-ipo-is-a-game-changer-for-the-saudi-economy/

International Monetary Fund (2019) Saudi Arabia: Selected issues. Country Report No. 19/291. https://www.imf.org/en/Publications/CR/Issues/2019/09/09/Saudi-Arabia-Selected-Issues-48660

Kerr S (2019) Saudi sovereign wealth fund nears $10bn loan. Financial Times, 11 July. https://www.ft.com/content/50df766e-a3de-11e9-974c-ad1c6ab5efd1

Kingdom of Saudi Arabia. (2017). Our vision. https://www.vision2030.gov.sa/sites/default/files/report/Saudi_Vision2030_EN_2017.pdf

Kirkpatrick D (2019) What to know about Prince Mohammed bin Zayed, the Arab ruler swaying Trump. New York Times, 2 June. https://www.nytimes.com/2019/06/02/world/middleeast/prince-mohammed-bin-zayed.html
Lillywhite (2015) Saudi Arabia: Female politicians and voters react after atleast 17 women elected in landmark poll. International Business Times, 14 Dec. https://www.ibtimes.co.uk/saudi-arabia-female-politicians-voters-react-after-least-17-women-elected-landmark-poll-1533285
Martin M (2020) Saudi wealth fund aims for more than 1,000 employees by year end. Bloomberg, 11 Feb. https://www.bloomberg.com/news/articles/2020-02-11/saudi-wealth-fund-aims-for-more-than-1-000-employees-by-year-end
Mazzetti M (2016) Saudi Arabia warns of economic fallout if Congress passes 9/11 bill. New York Times, 15 Apr. https://www.nytimes.com/2016/04/16/world/middleeast/saudi-arabia-warns-ofeconomic-fallout-if-congress-passes-9-11-bill.html
Merzaban D (2009) Saudi $5 billion investment firm eyes Europe, Asia, U.S. Reuters, 31 May. https://www.reuters.com/article/businesspro-us-sanabil-investment-interv/saudi-5-billion-investment-firm-eyes-europe-asia-u-s-idUSTRE54U0LS20090531
Mogielnicki R (2019) Expectation gap clouds Saudi Arabia's investment climate. The Arab Gulf States Institute in Washington. https://agsiw.org/expectation-gap-clouds-saudi-arabias-investment-climate/
Monk A (2010) Saudi's new SWFs are "stuck". Oxford SWF Project. https://oxfordswf.wordpress.com/2010/07/22/saudi's-new-swfs-are-"stuck"/
Onley J (2007) The Arabian frontiers of the British Raj: Merchants, rulers and the British in the nineteenth century gulf. Oxford UP, Oxford
Onley J, Khalaf S (2006) Shaikhly authority in the pre-oil Gulf: A historical-anthropological study. History and Anthropology 17(3):189–208
Rahman F (2019) Masdar eyes new renewable projects in Saudi Arabia. Gulf News, 16 Jan. https://gulfnews.com/business/energy/masdar-eyes-new-renewable-projects-in-saudi-arabia-1.61477000
Rashad M (2018) Saudi crown prince says sovereign wealth fund will surpass $600 billion target by 2020. Reuters, 5 Oct. https://www.reuters.com/article/us-saudi-pif/saudi-crown-prince-says-sovereign-wealth-fund-will-surpass-600-billion-target-by-2020-idUSKCN1MF2KC
Rashad M, El Gamal R, Al Sayegh H (2019) Saudi Aramco to buy SABIC in $69 billion chemicals megadeal. Reuters, 27 Mar. https://www.reuters.com/article/us-aramco-sabic-m-a/saudi-aramco-to-buy-sabic-in-69-billion-chemicals-megadeal-idUSKCN1R81QU
Raval A, Massoudi A (2019) Saudi Arabia replaces Aramco chairman with MBS ally. Financial Times, 2 Sep. https://www.ft.com/content/a8792b40-cd8c-11e9-99a4-b5ded7a7fe3f

Raval A, England A, Massoudi A, Kerr S (2019) The Aramco reshuffle shows the rise of Saudi wealth fund. Financial Times, 3 Sept. https://www.ft.com/content/85f17b32-ce12-11e9-99a4-b5ded7a7fe3f

Roll S (2019) A Sovereign Wealth Fund for the Prince. SWP Research Paper 2019/RP. https://www.swp-berlin.org/10.18449/2019RP08/#_Ref10701426

Seznec J (2020) The Saudi Aramco-SABIC merger: How acquiring SABIC fits into Aramco's long-term diversification strategy. https://www.atlanticcouncil.org/in-depth-research-reports/issue-brief/the-saudi-aramco-sabic-merger-how-acquiring-sabic-fits-into-aramcos-long-term-diversification-strategy/

Waterson J (2018) Saudi Arabia pays UK firms millions to boost image. The Guardian, 19 Oct. https://www.theguardian.com/world/2018/oct/19/saudi-arabia-pays-uk-firms-millions-to-boost-image

The Timor-Leste Petroleum Fund: From Buying Peace to White Elephants

Jerry Courvisanos and Anita Doraisami

Introduction

Despite the lack of political conflict over the establishment of an ideal Sovereign Wealth Fund to protect oil and gas revenue in Timor-Leste from the "Resource Curse," the story that unfolds in this chapter is one of political struggle over the Fund and its economic future. This struggle is not a traditional one between opposing parliamentary parties, but instead is between the Parliament of Timor-Leste (driven by the ruling governing executive) that supports "big development" and a vibrant strong civil society that espouses local-based sustainable grass-roots development.

This story unfolds around three key themes. The first of these are the crucial forces that propelled the creation of the Timor-Leste Petroleum Fund (referred to in this chapter as the "Fund"), with the blessing and

The original version of this chapter was revised: Author name and Author group order have been corrected. The correction to this chapter is available at https://doi.org/10.1007/978-3-030-78251-1_12

J. Courvisanos · A. Doraisami (✉)
Federation University Australia, Ballarat, VIC, Australia
e-mail: a.doraisami@federation.edu.au

E. Okpanachi and R. Chowdhari Tremblay (eds.), *The Political Economy of Natural Resource Funds*, International Political Economy Series,
https://doi.org/10.1007/978-3-030-78251-1_10

goodwill of the world community that was dismayed over the tragic circumstances behind the nation's founding. The second are the nascent fissures that arose with the conflict and compromises that developed in response to the administration of the Fund. The third is the problematic nature of the Fund's political and economic outcomes, arising from the changes in the Petroleum Fund Law, the magnitude and direction of expenditures from the Fund, and the questions as to whether the Fund has achieved its stated aims. The story is in its third act, after the euphoric creation and slow undermining of the Fund, and has only relatively recently moved toward a shift in the rules of the game. Thus, any conclusion can only be provisional as the third act is only unfolding now, but the signs indicate that it is a long way from its promising start.

Background

After the violence committed by the incumbent governing military for 23 years and immediately following a strong pro-independence referendum vote in August 1999, the United Nations Transitional Administration (called UNTAET) entered to prepare the region for nationhood (Kingsbury 2012; Leach 2017). On 20 May 2002, sovereignty was transferred to the first elected Parliament and government of the nation of Timor-Leste (Portuguese for "East Timor"). As Secretary-General of the FRETILIN Party, which had received a large majority of the vote in the first parliamentary elections, Mari Alkatiri became the first prime minister, while the President was Xanana Gusmão, the former guerrilla leader. During the transitional administration, the idea of a Sovereign Wealth Fund to protect against the Resource Curse became the subject of serious discussion. The United Nations considered the Norwegian model the appropriate exemplar to introduce before any oil fields came into operation (NRGI 2013). Such a fund enabled long-term development in light of many demands for immediate spending from such "windfall gains." These demands were massive, due to the extensive physical destruction prior to independence, with per capita income estimated to be about US$300 in 2002, and most development indicators extremely low. After the international contributions to aid reconstruction, incomes rose to about US$430 per capita in 2003. The population at the time, of approximately one million, still required substantial infrastructure and social development investments (Bacon and Tordo 2006).

Three separate oil or gas fields were identified in the Timor Sea between Australia and Timor-Leste, in which Bayu-Undan started production in 2004. Revenue from Bayu-Undan accrued 90% to Timor-Leste and 10% to Australia. By the fiscal year 2004–2005, revenues from this field were US$129 million, compared to spending from the central government account of US$75 million and GDP of less than US$350 million. Bayu-Undan production will cease by 2023 as reserves deplete (Bacon and Tordo 2006).

Hamadi and Gomes (2013) estimate that between 70 and 80% of the population is dependent on subsistence agriculture with low marginal productivity, output, and income, and significant exposure to food insecurity. Lundahl and Sjöholm (2013) point to a large number of improvements that need to be made in the agricultural sector, including introduction of new seeds, farming equipment, irrigation, transport to market, extension services, access to fertilizers and pesticides, and storage facilities. Both the lack of credit and the complexities of land titles stymie shifting from subsistence to sustainable agriculture.

In the longer term, surplus labor currently in subsistence needs to move to other advanced sectors to achieve higher incomes; however, these other sectors of the economy have shown no signs of employment opportunities emerging. Lundahl and Sjöholm (2009) estimate that the calculation of the annual growth of output required in these modernized sectors just to keep the share of the labor force in agriculture constant is 21% if the public sector is included, and a staggering 106% if not. Given this scenario, improvements in the agricultural sector itself are crucial for reducing poverty and promoting sustainable and more equitable economic growth in Timor-Leste.

Pre-establishment Stage

Norwegian experts provided input on a public consultation paper on the establishment of a fund, which Prime Minister Alkatiri issued in October 2004 in the face of the need to involve the whole population in the debate over its nature and use (Office of the Prime Minister 2004). The paper was supported by an extensive public education campaign that also solicited comments. It outlined the broad principles of the fund that the FRETILIN Government recommended for adoption in terms of its design and implementation. In February 2005, a draft act was published, inviting comments that were made public on the Ministry of Finance's website. The World Bank, the International Monetary Fund (IMF), several major Non-Government Organizations (NGOs), and individuals made extensive comments on the draft. The final version was unanimously passed Parliament on 20 June 2005: "This was the first occasion on which an Act had been passed unanimously by the Parliament of

Timor-Leste, thereby symbolising parliamentary unity on this important issue of petroleum revenue management" (Drysdale 2007, p. 84).

Thus, the Fund became law No. 9/2005 on 3 August 2005 (NRGI 2013). The political and community-wide support for this law was evident in that it was passed almost five years after the first petroleum royalty payment accrued to this fledgling nation after the arrival of UNTAET (Drysdale 2007). The aim of the Fund is to collect royalties from oil exploitation in an orderly manner so as to buffer (stabilize) the budget from volatile oil revenues, as well as reinvest such revenues in longer-term economic activities that do not depend on nonrenewable resource extraction (NRGI 2013; La'o Hamutuk 2018). Truman's (2008) review of 32 Sovereign Wealth Funds ranked Timor-Leste's Fund third, behind those of New Zealand and Norway, based on its structure, governance, transparency, accountability, and behavior.

The politics of Timor-Leste in the period prior to the establishment of the law were unique. Although the Portuguese colonial administration did know of oil resources in Timor-Leste and exploited it for local use, large exploitation for export began only in 1989 with the signing of the Timor Gap Treaty between Australia and Indonesia. This treaty adopted a maritime border and identified exploitation zones between the two countries. Woodside Petroleum (Australia) and Phillips (now ConocoPhillips, USA)—along with another 24 companies—opened the first oil fields in the early 1990s (Dal Poz 2018). This treaty automatically terminated when Indonesia no longer controlled the waters south of what was then the East Timor UN administrative region. On 10 February 2000, Australia and UNTAET reached an agreement on the continued operation of oil and gas exploitation under the same operating rules as those governing the Timor Gap Treaty. All royalties would remain in trust until the new nation was established and developed a form of agreement on the management of those funds, which occurred with the passing of law No. 9 (Hendrapati 2018). As such, there were no powerful interests existing in Timor-Leste that could suffer from any new management fund law (Anderson 2013). With the broad community consultative process outlined above, the strong international goodwill to ensure the best funds operation, and no local or foreign negative commercial and monetary implications, the establishment of the Fund lacked any fragmentation of views and conflict did not arise (Dal Poz 2018).

Unverifiable rumors circulated that a few leading individuals in the independence struggle at the time objected privately to the idea of a

Sovereign Wealth Fund safeguarding the nation from the Resource Curse. The minority view supported "big development" based on maximizing the nation's economic activity (i.e., GDP) quickly with the standard neoclassical economics "trickle down" effect to eventually benefit all in the country (Courvisanos 2012, pp. 14–16). However, given the above concentration of social, economic, and political interests in establishing such an ideal fund, this alternative "big spending" view did not emerge publicly in the pre-establishment stage, but did appear in the later operations stage, as will be outlined later in this chapter.

Establishment Stage

The Fund established an institution with strong resilient rules and structures to manage the petroleum revenue, while the Petroleum Activities Law enables Timor-Leste to contract with oil companies for exploration and production that provides the Fund's revenue. This legal institution was tested during the 2006–2008 civil crisis when "the lack of bridging social capital" led to "strong resentment of the political leaders and a very large divide between the decision-makers and the community" (Drysdale 2008, p. 57). The crisis began between elements in the military over discrimination, and culminated in the involvement of the army and the police, as well as a coup attempt. However, because the Fund did not become a focus of these grievances, the mechanisms to manage petroleum revenue were not subverted; as well, two successful international licensing rounds for new petroleum contracts were conducted during this "crisis."

Essentially, this strong institution is due to the ideal base of the Norwegian model and the wide consultation process in the pre-establishment stage that ensured agreement on two goals that underpin payments out of the Fund. The first of these goals, set in legal principle, is that all payments out are made to the state budget and cannot exceed a ceiling set by Parliament at the time of budget approval. They will generally be equal to that amount needed to cover the non-oil deficit in the planned budget, while the residue will then remain in the Fund. The second goal is an assurance that the Fund will pay out over its lifetime an amount to finance the state budgets that is viable indefinitely, as shown in a formula in the Act (Bacon and Tordo 2006).

These two goals can be inconsistent, because the sustainable income from all sources need not be equal to the non-oil budget deficit. The law specifies that if this occurs and the proposed non-oil budget deficit is larger than sustainable income, then a Petroleum Fund Consultative Council (PFCC) should be set up, with civil society organizations represented, to advise Parliament. The PFCC reports on the potential overspending issue to Parliament, which will then make the final decision. In this way, the actual rules governing payments out of the Fund permit a compromise between ensuring that the real wealth of the nation stays intact and using oil revenue for immediate budgetary support.

The Estimated Sustainable Income (ESI) from petroleum revenue is related to the discounted sum of expected future revenue. The Act makes no mention of how the price assumptions concerning future expected revenues from oil/gas are to be calculated, unlike the cases of Chile or São Tomé and Príncipe (Bacon and Tordo 2006). The PFCC advises on whether planned withdrawals are in excess of expected ESI. Thus, the primary duty of the PFCC is to see that this calculation is made. Without guidance on the principles to be followed, this use of "excess" funds could prove controversial, especially because it involves assessments of unknowable future oil prices and must also make allowances for any quality in public spending which ensures the real wealth of the nation stays intact, and using the revenue for immediate emergency budgetary support.

The FRETILIN-led first Alkatiri Government maintained its resolve to pursue cautious and autonomous fiscal policy, carefully managing small budgets and, despite the lack of resources, avoiding debt. In education and health, this government built on the country's constitutional commitment to education and health as rights rather than as commodities. In this sense, the 2002 constitution had established, and the government maintained, some "human development" themes, as opposed to the "pay for service" ethos of market economies (Anderson 2013, p. 228). The government began with strong commitments, pledging "more than 35% of resources" to education and health in its first budget. However, it delivered only 27 and 25% in the 2004–2005 and 2005–2006 budgets, respectively (Anderson 2010, p. 32).

Scambary (2015) contends that the struggle for independence generated high expectations of a peace dividend, especially among former guerrillas, whose sense of grievance was compounded by exclusion from employment opportunities in the newly created security forces. The demobilized veterans mostly missed out on jobs in the civil service and the new police force, and many faced unemployment and significant hardship. Consequently, organized veterans groups began to conduct parades and operate as political lobby groups, and were perceived as a threat

(World Bank 2008). This threat came into sharp relief with the civil crisis mounting and the disgruntled military attempting a coup. Alkatiri was forced to resign as prime minister in August 2006. Violence from this crisis lasted for almost two years, with the involvement of veteran groups unhappy with their treatment since the withdrawal of the Indonesians (Scambary 2015).

In June 2007, elections were held and an alliance of three smaller parties called the Parliamentary Majority Alliance (AMP) led by former President Gusmão won government, with Gusmão becoming prime minister. AMP campaigned on the premise that it would manage Timor-Leste's petroleum resources differently; Gusmão stated that he would unlock the revenues of the Fund, suggesting that withdrawals would exceed those of the previous FRETILIN Government (Drysdale 2008; Anderson 2013). This marked the beginning of economic and political strains in what had previously been a conflict-free zone in the politics of Timor-Leste.

The 2008 State Budget was the first budget of the AMP-led Gusmão Government. The government proposed, and Parliament approved, a mid-year budget rectification that increased the 2008 budget by 126% to nearly US$788.3 million with a transfer of $US290.7 million above the sustainable level from the Petroleum Fund (La'o Hamutuk 2009). The largest expenditure item in the 2008 mid-year budget update was the setting up of an "Economic Stabilization Fund" worth US$240 million. In that same year, further overspending of the Fund, due to the setting up of the Economic Stabilization Fund (ESF), was blocked when the courts found it unconstitutional (Wallis 2014). Since then, the challenge has been to maintain the integrity of the Fund and its goals under circumstances in which the ruling government of the day arranged to support "big development" with vast public spending on large infrastructural projects (Anderson 2013).

According to the Ministry of Finance (2008), the ESF sought to stabilize prices through market intervention by subsidies to enable the supply of critical commodities, such as food and construction materials, at affordable prices. As Doraisami (2009) notes, at the time there was little information provided on the conduct of this market intervention and the cost of these subsidies to the public purse. In contrast to the 30 per cent of the 2008 Budget allocated to the ESF, approximately 3.98 per cent was allocated to the agricultural sector in which 70–80 per cent of the population are dependent. The same proportion was allocated to the health sector, while education allocation amounted to 6.64%.

The strain on the Fund, given the government's commitment to big development, resulted in eventual amendments to the Fund in August 2011 that altered investment policy. This policy change would increase returns at the cost of increased investment risk, while stretching spending well beyond the ESI. Then, a further undermining of the Fund occurred in changes to the Act and the underlying Petroleum Activity Law in 2019. These changes and their implications for the operations of the Fund are discussed in the next section.

Operations

Operations of the Fund were altered twice since its inception, when Parliament changed the rules under which the Fund functioned. This section discusses the operations of the Fund in each of the three distinct rule periods.

Sustainable Management Under Stress: July 2005–August 2011

The first period, the original form of the Fund, lasted from its inception in July 2005 to the first set of amendments in August 2011. The setup during this period provided the foundations for the Fund's operations on an enduring basis. The operational management of the Fund during this period, as described in this section, has been sourced from the Timor-Leste Government (2005, 2009, 2010) and Ministry of Finance (2011), as well as from Anderson (2010), Jensen (2010), and McKechnie (2013).

To give effect to the two basic goals of the Fund, the ESI formulation consists of three operating mechanisms. The first is a governance structure, at the center of which is the Finance Minister, the Investment Advisory Board (IAB), and the Banking and Payments Authority (BPA) which was transformed into the central bank, Banco Central de Timor-Leste (BCTL). The central bank manages the Fund pursuant to an agreement with the Minister, while the IAB is mandated to provide expert external advice. The BCTL, with the Minister's approval, may appoint one or more external investment managers. The PFCC is charged with advising the Parliament; it may also hire if it so chooses.

The second operating mechanism is the regime (principally Articles 14–15) set up for the investment of all petroleum-and-gas-related revenues that accumulate in the Fund. This law allows specific classes of investments to be held by the Fund as "qualifying instruments." These are US dollar-denominated fixed interest bonds and other "debt instruments," which must account for "not less than" 90% of the Fund's revenues. These "qualifying instruments" can only be from issuances of highly rated States or banks. The other "not more than" 10% of revenues can be invested with fewer restrictions. The intent of this provision is to ensure financial prudence in investments in order to minimize risk.

The third operating mechanism is a regulated process (mainly Articles 7–9) for withdrawal of revenues for use in government expenditure. Debits from the Fund can only occur via electronic transfer to "a single State Budget account" after the relevant budget law is approved by Parliament and promulgated. A central prerequisite for this process is that an audited statement of the ESI of the Fund is presented to the Parliament. If any amount over the ESI is to be drawn down by the government, the Parliament must be provided with a "detailed explanation of why it is in the long-term interests" of the country to do this (Article 9). There are no particular criteria set out for such over-ESI spending. The ESI is based on the Norwegian model, with withdrawals from the Fund guided by the real rate of return (RRR) on the total sovereign wealth. This approach has the benefit of being intuitively simple, appealing to the notion of intergenerational equity,and offering a stable source of income to the budget (IMF 2005).

The ESI for any given fiscal year is defined as the maximum amount that can be appropriated and transferred from the Fund to the budget while retaining sufficient resources to appropriate an equal amount in all subsequent years. While the ESI can be seen as a fiscal rule, it does not have a rigid ceiling. The law stipulates that actual withdrawal may exceed the ESI as long as it is justified by government and approved by Parliament.

The ESI for a given fiscal year is calculated according to the following formula:

$$\mathrm{ESI} = r[V + \sum_{t=0}^{n} \frac{Rt}{(1+i)t}]$$

where:

- r: specified RRR of 3%;
- V: balance of the Fund at the start of the fiscal year;
- Rt: budget projection for revenue in year t;

- *n*: last year of projected receipts; and
- *i*: discount factor specified as the nominal yield on US government securities averaged over years 0 to *n*.

The Fund Law states that all assumptions upon which the ESI calculation is based shall be prudent, reflecting international best practices and internationally recognized standards which are certified by an independent auditor (Timor-Leste Government 2005).

In its original form, 90% of the Fund resources were to be invested in investment-grade US dollar debt instruments, such as bonds, rated at least AA-by Standard & Poor's or Aa3 by Moody's. The remaining 10% can be invested in instruments that are issued abroad, liquid, transparent, and traded in markets of the highest regulatory standard (IMF 2009). This 10% design for formulation aims to build expertise in managing risky assets and to access higher-return investments. This investment strategy enabled the Fund to weather the 2008–2009 Global Financial Crisis (GFC) well (Drysdale 2012).

Over the first period (2005–2011), the Fund's 90% investment strategy yielded an RRR of approximately 2%, which was below the 3% assumed in the ESI formulation (Jensen 2010, p. 35), but the only Sovereign Wealth Fund in the world which did not lose money during the GFC (Scheiner 2020a). The problem this created is that if the conservative investment requirements in the law were retained, and the 3% assumption revised downward, the ESI would fall considerably. This would also require a commensurate decrease in government spending, if spending were to remain inside the ESI formulation. Alternatively, if the rigid investment restrictions were loosened to allow a larger portion of the Fund to be invested in higher-yield instruments, the probability of achieving the 3% annual return that the ESI assumed would be greatly increased. This purely technical dilemma manifested itself as a political debate on the nature of economic development in the country.

After the resignation of Prime Minister Alkatiri, new elections eventually resulted in the government of Prime Minister Xanana Gusmão taking office in August 2007. It faced a daunting task of maintaining order and stability: the economy in 2006 had contracted, with a decrease of 5.8% of real non-oil GDP growth. Approximately 150,000 displaced people were scattered around the country and in temporary camps around Dili. The proportion of unemployed youth engaging in gang activities had increased dramatically. The Gusmão Government addressed this political and social concern, first in 2008 with new social assistance measures, then in 2009 with a more thoroughgoing package (Porter and Rab 2010).

Of the 2008 social assistance programs, the largest was a pension to veterans of the independence struggle. The second largest was funds to significant constituencies impacted by the civil crisis—displaced persons, dissident defense "petitioners," and elderly or disabled pensioners. The third, and by far smallest, program was a conditional cash transfer benefit to single mothers with children enrolled in school. This latter program, called *Bolsa da Mãe* ("Mother's Purse"), was based on a similar program in Brazil. School feeding, cash-for-work, rice importation, and food subsidy completed all the 2008 measures (World Bank 2013). The poverty impact of these programs was assessed by the World Bank (2013) as not commensurate with the level of spending, with 60% of the bottom two quintiles not reached by any of these social assistance programs, because these measures were political responses by a new government to a civil crisis rather than a poverty reduction program. The lion's share of the 2007 measures is devoted to veterans' payments, which is a very high income for the recipient, estimated at over 426% of the per capita non-oil GDP. The payments are made to veterans from the independence struggle regardless of their level of income, even though these veterans form only 1% of the population. Veterans' payments exceed the entire country's health budget and about 80% of the education budget (Kent and Kinsella 2015).

On 22 August 2009, the Gusmão Government informed Parliament that the heavy oil project (diesel power plants and electric grid) was running into problems, and that $70 million of the $85 million appropriated for it during 2009 was to be reallocated into the Pakote Referendum Package. The World Bank (2014) described this package—based on government input—as aiming to stimulate rural development by creating a local entrepreneurial class through a modality that funded contractor capitalization and capacity and allow Timorese companies to win government contracts, thus generating employment and delivering high-quality infrastructure. On the other hand, the independent think-tank La'o Hamutuk (2009) described it as "a textbook case of Resource Curse thinking, [being] wasteful, uncontrolled, impetuous spending unrelated to the plans in the 2009 Budget."

The 2008 and 2009 initiatives must be viewed as responses crafted in a post-conflict environment to "buy" peace. However, they have had long-lasting consequences for government budgets going into the future, raising viability concerns for the Fund and serious implications for Timor-Leste's future prosperity (Doraisami 2018). Akmeemana and Porter (2015) assert that public spending was the only means available to stabilize the country and bring "into the tent" all political opponents and potential opponents with the capacity to mobilize violence. The "tent" involved significant rewards for these various actors, calling it an "independence" dividend to the broad populace; however, this dividend affected the viability of the Fund. To "pay" the dividend in 2008, the new Gusmão Government made the maximum allowable withdrawals from the Fund according to the ESI formula, then exceeded this formula after the 2009 package and this excess continued to grow until 2012 (see Table 1). This widening excess over ESI raised long-term legal funding issues that resulted in the Gusmão Government seeking major amendments to the law.

Infrastructure Investment Management Undermining Rationale: September 2011–September 2018

Substantial amendments to the Fund law occurred in September 2011, which altered the design of the Fund in order to earn higher financial returns by allowing greater flexibility with a more diverse portfolio. Article 15 amended the law to allow up to 50% of the Fund to be used for investment in listed shares traded in regulated foreign financial markets. The amendments also allow up to 5% of the Fund to be placed in other investments, including financial derivatives, with the approval of that investment asset class by the Finance Minister and provision of information to Parliament. The amended law also permits the use of up to 10% of Fund assets as collateral for public borrowing within the government's debt management system, with proper accounting for resulting contingent liabilities in the Fund's financial statements. No less than 50% of the Fund would still be in bank deposits or investment-grade debt instruments. While BCTL remains the operational manager of the Fund, the amended law allows the Finance Minister to determine the manager with the advice of the IAB.

An examination of Table 1 on the state of the Fund between 2008 and 2019 indicates its notable change in character over that decade. After 2008, withdrawals were always in excess of the legal maximum allowed by ESI formulation under the law. This excess rose significantly through to 2011, even discounting for the manipulation during these years of the allowable ESI with the government opting for a less

Table 1 The Timor-Leste Petroleum Fund: selected indicators 2008–2019

	2008	*2009*	*2010*	*2011*	*2012*	*2013*	*2014*	*2015*	*2016*	*2017*	*2018*	*2019*
Estimated Sustainable Income (ESI)	396	408	502	734	665	787	632	639	545	482	550	529
Annual withdrawals/ESI (%)	–	125	161	143	224	092	115	200	228	224	178	183
Annual total withdrawals	396	512	811	1055	1495	730	732	1279	1245	1079	982	969
Petroleum fund balance	4197	5377	6904	9310	11,775	14,952	16,539	16,218	15,844	16,799	15,804	17,692
Production (mmBoE)	71	69	63	69	74	65	50	52	49	41	37	38
Crude oil price	97	62	79	111	112	109	99	52	44	54	71	64

Note All figures are in millions of US dollars
Source IMF (2012, 2013, 2016)

conservative oil price estimate being placed into the ESI calculation, despite actual recorded lower crude oil prices. Withdrawals were lower in 2013–2014 because the government had withdrawn much more than it could spend in 2012. With the Fund balance rising strongly until 2014, the Gusmão Government and Parliament were confident that the Fund would be financially viable, emboldening them to make even larger public expenditures, particularly infrastructure, as seen in Table 2.

The oil/gas sector accounted for about 70% of GDP and almost 90% of total government revenue between 2010 and 2015, with the Bayu-Undan and Kitan fields the main sources of this petroleum revenue. Two factors militate against the viability of the Fund. One is the much lower crude oil price in the latter years of Table 1, undermining the government euphoria over large price gains in 2012–2013. The other factor is the much lower levels of production and the projection that existing oil fields will be depleted by 2023, with no new fields currently being developed or

Table 2 Total budget executed expenditure 2008–2019

	2008	*2012*	*2013*	*2014*	*2015*	*2016*	*2017*	*2018*	*2019*
Total executed expenditure	483.8	1195.1	1050.2	1383.0	1337.2	1636.7	1224.6	1157.0	1235.5
Executed recurrent expenditure	355.9	691.3	728.2	895.5	916.4	857.1	821.8	826.9	872.9
Salaries and wages	50.3	130.9	141.8	162.5	173.4	179.8	195.7	192.9	203.2
Goods and services	221.3	309.7	354.3	428.0	392.5	349.4	327.0	308.9	357.5
Total public transfers	84.3	218.5	197.1	291.7	432.6	471.1	400.4	312.5	339.2
...of which veterans' payments	3.6	109.7	73.4	60.9	119.9	97.7	94.6	91.4	91.1
Capital expenditures	127.9	503.8	352.0	465.5	306.6	602.7	279.5	330.7	315.6
Domestic revenue	69.4	137.7	151.8	168.0	170.1	198.7	182.8	190.7	185.6
Loans	–	–	8.8	15.7	23	30.6	30	39.2	49.7

Note All figures are in millions of US dollars (projected)
Source La'o Hamutuk (2018)

even on at the level of planned field development with equity and finance in place (La'o Hamutuk 2018).

As Table 2 illustrates, between 2008 and 2019 government expenditures have nearly tripled, while transfer payments quadrupled and infrastructure spending almost tripled. Timor-Leste is an outlier both in the region and for its level of per capita GDP as far as transfer payment expenditures are concerned (World Bank 2013). However, budget appropriations have always exceeded expenditures as the government has faced continuing problems with budget execution of infrastructure (Doraisami 2018).

The viability of the Fund must be evaluated against the major infrastructure projects that are draining the Fund: the national electricity grid, the special economic zone in the Oecusse enclave (ZEESM), and the petroleum resources development corridor on the remote south coast, known as Tasi Mane. Electrifying the whole mountainous country is a massive effort that has been virtually completed. A major criticism of this project is the expensive and outdated technology of centralized oil-fired power plants, with more than a quarter of the population who live in remote areas not expected to benefit from it (La'o Hamutuk 2017). There have also been reports of corruption in the grid construction contracts, and the continuing unreliability of the electricity supply (Scambary 2017), the benefits of lighting and power have had extensive multiplier effects in the long term, where "[h]ealth, education, agriculture, business and economic progress draw their vigour and higher productivity from the power of electricity" (Pereira 2015, p. 24). The ZEESM project, which aimed to facilitate foreign investment, envisages an airport equipped to land jumbo jets, multistory hotels, a world-class university, a model hospital for medical tourism, and features meant to earn Oecusse a "green city" designation. In 30 years, the plan is to spend US$4.11 billion, with US$1.36 billion drawn from public funds and the remainder from the private sector (La'o Hamutuk 2015). Scheiner (2015) states that loans from foreign institutions will incur hundreds of millions of dollars in debt, to be repaid well after petroleum revenues have ceased. Tasi Mane is the largest infrastructure development, expected to account for 15–20% of all infrastructure costs over the 2017–2020 period (De Haan 2017). The government's goal for Tasi Mane is to bring value-added development to Timor-Leste's south coast with a pipeline, gas processor, oil refinery, and supply base, providing direct economic benefit through downstream petroleum industry activities (Timor-Leste Government 2011; Timor Gap 2015). The latter two projects are expected to consume more than a quarter of the infrastructure budget through to the early 2020s (Doraisami 2018).

The sustainability of the Fund was being undermined by excess spending over ESI, declining Fund balance, continuing low oil prices and gas production and prices, with rising loan commitments (La'o Hamutuk 2018). Doraisami (2018, p. 254) argues that the excess spending is due to "[f]ront loading infrastructure and high expenditures on cash transfers [which] has squeezed out spending on education, health and agriculture." These concerns raised by independent researchers can be juxtaposed with the Parliament's ongoing support of the Gusmão Government's public spending priorities. The FRETILIN Party, still led by Alkatiri, has remained in opposition since its demise in 2006, except for a one-year (2017–2018) period in which Parliament refused to approve the government's programs and budget. Since 2006 FRETILIN has provided only tepid opposition, either being dragged into a "national unity" government led by Gusmão or having a FRETILIN President veto bills that go back to Parliament and are then passed, or sometimes marginally altered. Both Scambary (2015) and Doraisami (2018) describe the large infrastructure projects as "white elephants" that, together with poorly designed cash social payment transfers, undermine the Fund's intended purpose of avoiding the Resources Curse. Scambary (2015) contends that the investment projects are characterized by negative social surplus, and selected either on the basis of servicing clientelist networks or political prestige, with little regard for likely economic and financial return.

Equity Investment Management Under New Economic Order: From January 2019

The resolution of the maritime border issue with Australia on 6 March 2018, which came into effect on 30 August 2019, raised serious prospects for the development of a very large oil/gas field known as Greater Sunrise, with the pipeline and downstream processing via the Tasi Mane project. This is a key part of the Gusmão Government's development strategy as it expects Tasi Mane to contribute broadly to the economic development of Timor-Leste and the specific regional development of the isolated south coast, generating over 10,000 direct jobs and more than 50,000 indirect jobs (ACIL 2016). Under the Australia–Timor-Leste Maritime

Border Treaty, if processing occurs in Timor, the new division of upstream revenues under the treaty will allocate 70% to Timor and 30% to Australia, compared with the previous 50–50 split. Stewart (2018) reports that at current prices, this would mean that Timor's share over the next 35 years would rise from US$20 billion to US$28 billion, while Australia's share would fall from US$20 billion to around US$12 billion. If, instead, the gas was to be developed in Darwin, Timor would be compensated by receiving a higher share of the revenue, at 80%, with 20% for Australia. With Timor-Leste rejecting the Darwin option, this gas processing plant has been sold to service another gas field.

Two of the initial Greater Sunrise private partners preferred the less risky oil/gas pipeline and processing in Darwin. This preference propelled Xanana Gusmão to appear on public TV on 1 October 2018 to announce that the government would purchase the equity shares of these two shareholders in the Greater Sunrise gas field. The government reached agreements with ConocoPhillips (2 October 2018) and Shell Australia (21 November 2018) to acquire their respective shares in the joint venture for a combined US$650 million. This would give the government a 56.6% majority share in the joint venture, with Woodside and Osaka gas receiving shares of 33.4 and 10%, respectively. The government hopes that this will remove key obstacles to developing the field and the onshore Tasi Mane project. However, because this equity investment required funds, the Fund serves as the equity management vehicle.

To facilitate the Fund's financing of the equity purchase, the Petroleum Activities Law was amended in January 2019 to allow the State to participate in equity investment. Parliament promulgated amendments to the law on Petroleum Activities as Law no. 1/2019 on 18 January 2019, retroactive to 27 September 2018, despite an initial veto by the President. Another amendment enabled the Fund to invest directly in the national oil company, TIMOR GAP. The Ministry of Finance was then able to revise the Fund's investment policy and instruct the Banco Central de Timor-Leste to reduce the proportion allocated to international stocks from 40 to 35% and include investment in TIMOR GAP under "other eligible investments" (which have a 5% cap). With these resources, TIMOR GAP purchased the shares of ConocoPhillips (30%) and Shell (26.6%) in the Greater Sunrise fields. Timor GAP will pay 4.5% annual interest on the $650 million loan, with an 8-year grace period and an 18-year repayment term. This amendment law includes removal of the 20% limit on State participation in petroleum projects and the need for prior inspection by the Chamber of Auditors for any petroleum-related contracts. TIMOR GAP estimates that the Greater Sunrise fields could generate $22 billion in profit over a multi-decade project life (World Bank 2019; La'o Hamutuk 2019a).

Developing the Greater Sunrise fields will be a huge task for a small improvised country. The fields are estimated to hold between US$40 and $60 billion worth of oil/gas reserves, but investment costs—which include both capital expenditure (capex) and operating expenditure (opex)—are also expected to be high. The delay in the resolution of the maritime border issue with Australia has exacerbated difficulties with Australia being accused of siphoning millions of dollars of Timor-Leste's oil revenue, thought to be in excess of the entire amount of aid it has given to Timor-Leste in foreign aid (Davidson 2019). TIMOR GAP has stated its intention to develop Greater Sunrise without using Petroleum Fund resources. Financing options for both onshore and offshore developments include project financing, private equity, joint venture, and international loans, among others. Securing private sector participation in these developments, coupled with adequate safeguards, would be critical to minimize commercial risks that are inherent to operations in the sector—owing to highly volatile prices and demand—and protect the Fund (World Bank 2019). The danger is that if commercial investors keep insisting that the Tasi Mane downstream plan is less profitable than other gas projects elsewhere, investors will walk away, the resources under the sea will benefit no one, and Timor will have no petroleum resources to exploit. On the other hand, Australia is neutral about whether the gas is developed in Timor or Darwin, so long as it maximizes the commercial potential.

With the equity-based changes to the legal structure of the Fund, from now on each year's government budget will need to be scrutinized for its fiscal stance as the Fund comes under more financial stress. The 2019 budget provides a marker for this yearly stress test. This budget allowed for equity shares to be financed largely by withdrawing US$969 million from the Fund, the largest withdrawal in its history, as only US$186 million was available from State revenue (La'o Hamutuk 2019b). A massive 50% of this budget was allocated to social payment transfers. This amount represents a significant increase in the fiscal deficit and, consequently, due to a lack of internal financing capacity, the State was obliged to withdraw from the Fund 249% above the ESI. As of March 2020, the second proposed 2020 budget was also rejected by Parliament. This has placed the country's public funding—including large extraordinary withdrawals to manage the emerging COVID-19 pandemic—under an interim "duodecimal" fiscal regime for several months which allows monthly appropriations of 1/12th of the previous (2019) budget (Scheiner 2020b; La'o Hamutuk 2020).

As of early 2020, oil revenues were decreasing as the volume of production from the Bayu-Undan field decreased significantly since it is almost depleted. This is intensified by continuing low (and very low under COVID-19) oil/gas prices. The pace of withdrawals from the Fund, with embedded excessive social payment transfers at very high levels and exacerbated by the pandemic, may never be replenished. This jeopardizes the viability of the Fund itself, accelerating the eventual extinction of the Fund (La'o Hamutuk 2019b; La'o Hamutuk 2020).

Natural Resources Funds (NRFs) in Operation: Political Struggles Over the Funds, Continuity, and Change

Economic development in Timor-Leste has reached a critical juncture, of which the Fund is a crucial element. At the inception of the Fund, everyone was in support of its basis in the Norwegian model of prudence and commitment to future generations. As well, the first government pursued a cautious "small development" policy that emphasized education and health and strove to avoid debt. Thus, the establishment of the Fund was marked by a noticeable lack of conflict, with compromise arising only at the edges in addressing minor amendments to the proposed Fund model.

Two strong forces emerged in the early years of the newborn nation that were diametrically opposed to the inception forces that led to tensions in the management of the Fund. The first of these was the political and civil crisis of 2006–2008 that saw the demise of the first "small development" government. The government of the time needed to "buy" peace with massive social payment transfers that were not linked in any way to poverty reduction. The second factor was the same government's adoption of a "big development" policy agenda based on the 2011–2030 Strategic Development Plan (Timor-Leste Government 2011), with the Fund being viewed as a "honeypot" that, as it grows, attracts more interest and provokes dreams of major infrastructure projects. Courvisanos and Boavida (2018)critically examine the SDP in the context of Timor-Leste's professed support for the United Nations Sustainable Development Goals (SDGs) and find the "big development" approach as inevitably clashing with the SDGs Agenda, which they evaluate as second-best to the SDP.

The most favorable aspect of the Fund that has continued since its inception is that oil and gas receipts are properly accounted for and investments are managed effectively with a coherent framework by local Timorese within the BCTL. Beyond the technical management of the Fund, the political management of the Fund is constantly challenged outside Parliament. As successive governments have more closely aligned with big development, there have been only a few voices of opposition within Parliament. There is a lack of a strong voice in Parliament for local-based sustainable grass-roots development that could provide the basis for critical political management of the Fund. Parliament often acquiesces to big spending governments as it recognizes the limited role it can play, and that any of its proposed policy changes are inevitably rejected by government, thus delaying any valuable ongoing programs (Leach 2015).

The issue over management of the Fund has raised significant concerns in the broader Timorese community. This is especially observed in the organizations of the well-educated civil society who express doubts about the viability of the Fund and its use for non-sustainable development at a time when fossil fuel prices are low and development is being phased out globally. This position has been strongly advocated by La'o Hamutuk, the civil society organization that monitors and analyzes the economic and social developmentof the country and places this information on its widely accessed website. La'o Hamutuk regularly raises the spectre of the threat of a Resource Curse despite the Fund's well-structured creation, and this is due to governments controlled by Gusmão undermining the goals of the Fund as described above. As Scheiner (2020b) of La'o Hamatuk asserts, renewable energy is getting cheaper, new offshore gas fields are being discovered around the world, and growing awareness of the climate emergency will make fossil fuels less lucrative in the future—all of which makes it harder to justify the Tasi Mane project.

Over time as governments have become more aggressive with withdrawals from the Fund, La'o Hamutuk's counsel has been less able to influence policy decisions or persuade Parliament to alter the approach, except for minor some amendments at the fringes of development. Like

La'o Hamutuk, other organizations in the country are fully supportive of the Fund and its technical management. The opposition has become more intense over the governance of the Fund as large infrastructure and equity-based projects are seen increasingly as unrealizable, but requiring greater Fund commitments. These organizations focus on specific concerns in relation to the big developmentstrategy by the government that is underpinned by the Fund. In a report to the United Nations Development Programme (UNDP), Courvisanos and Jain (2018) note spokespersons for various organizations setting out these concerns. One social enterprise stated, "we also have the Strategic Development Plan [SDP], these two things [SDGs and SDP] sometimes confuse me." Even more explicitly, a financial institution detailed its concern as follows:

> Development tends to focus in the city only and [the rest are] left behind...[yet] we don't pay attention to the agriculture sector, but only focusing on big/mega projects and big industry ... It is very important to make industry out of agriculture, as most of our population depends on the agricultural sector.

Another financial institution interviewee recognized the overwhelming major economic activity undertaken by the country's population is farming: "climate change impacts on agriculture production is very serious, as most of our farmers depend on agriculture for their main source of income and food." Furthermore, an NGO interviewee suggested an alternative model of development based on sustainability both for the ecology and the Fund: "Yes, we think this is very important as SDGs aim to find a model for development that will protect our environment and people, and to reduce the social gap." When focusing on oil development, an education-based civil society organization (whose spokesperson was a foundation member of the PFCC) questioned the development strategy:

> Today [when] people talk about economic growth in Timor, it is focused on oil. We have money because we have oil, but when oil is gone, will we still have economic growth? ...So now how can we build an economy on the basis of nature, harmony between the human, the economic and environment?

Along the same lines, an agricultural cooperative stated: "People in rural areas are left out and just watch the so-called development that does not benefit our people."

With all governments since 2006 using the Fund as the mechanism to undertake big developmentyinfrastructure, they have asked the many under-funded NGOs in the country to carry the weight of executing critical social, health, and ecological programs. This is problematic given the NGOs' lack of power to determine the overall sustainable development direction of the country and the declining commitment by international NGOs to the country as other trouble spots in the world demand more of their attention with their limited resources (Barrowman and Kumar 2018).

The FRETILIN President, Francisco Guterres (Lú-Olo), provides a voice of dissent, but one that is situated outside Parliament. He criticized the 2019 State Budget with its $2.1 billion expenditure, Timor-Leste's largest ever, for devoting too much to buying foreign oil assets. Thus he vetoed the Budget for four reasons that are a window to wider community disagreement: its financial unsustainability, its orientation that is contrary to the Constitution and structuring laws, its high budget imbalance, and the absence of alternative policies emphasizing education, health, and other public services (La'o Hamutuk 2019b). He noted that: "These values are so low that they do not meet the minimum requirements of social services and economic growth" (Raimundos 2019). Returned to Parliament following the veto, the Budget was passed after the $650 million to buy into Greater Sunrise was removed, with the opposition supporting the bill; instead, the purchase is being financed through borrowing from the Fund, adding increasing strain on its sustainability into the future. The budget that finally passed did somewhat modify spending levels, but did little to alter the ethos and substance behind the fiscal policy stance underlying the expenditures proposed, and exacerbated financial difficulties with the Fund.

Conclusion

This chapter sets out a three-act play whose first act was the euphoric creation of the Fund soon after it became a sovereign state. The Fund established an institution with strong resilient rules and structures to manage the petroleum revenue, with the Petroleum Activities Law underpinning the type of activities that the Fund could be involved with. In the second act, political realities in a post-conflict society became apparent and resulted in stresses on the viability of the Fund. In this period, the availability of Fund resources was tapped to buy the peace. The third act has only just begun, with the shifting of the Fund "rules of the game" that have the potential to undermine the Fund completely in a post-fossil fuel sustainable-development world.

The best aspect of the Fund is that oil and gas receipts are properly accounted for and investments are managed competently within a coherent framework. Perhaps the major achievement of the Fund is that it has provided transparency. The most concerning aspect of the Fund is its governance, which has eroded this transparency over time at a significant financial cost to the country. This is due to the "founding father" independence leaders still dreaming of a Western-style "big development" approach to addressing the poverty of Timor-Leste. This dream is based on Greater Sunrise providing a new dawn for fossil fuels by "draining" the Fund and paying it back with an extensive resource development industry. Despite this optimistic outlook, the signs indicate that the Resource Curse could plague Timor-Leste if large fossil fuel revenue is not realized, with no alternative clear pathway to a sustainable development future evident in its political landscape.

Acknowledgements The authors would like to thank Agostinho Maia (Banco Central de Timor-Leste) and Charles Scheiner (La'o Hamutuk) for their local-based input into the data underpinning the narrative, and for feedback on a draft version of this chapter.

References

ACIL [ACIL Allen Consulting] (2016) Tasi Mane project: Potential implications for the economy of Timor-Leste. Report to Timor Gap E.P. Brisbane: ACIL Allen, 30 Aug

Akmeemana S, Porter D (2015) Securing a new ordering of power in Timor-Leste: The role of subnational spending. In: Ingram S, Kent L, McWilliam A (eds) A new era? Timor-Leste after the UN. ANU Press, Canberra, pp 17–29

Anderson T (2010) The Petroleum Fund and development strategy in Timor: A report for Timor-Leste's Petroleum Fund Consultative Council. Resource document. Petroleum Fund Consultative Council. http://citeseerx.ist.psu.edu/viewdoc/download?doi=10.1.1.618.1409&rep=rep1&type=pdf. Accessed 10 Dec 2018

Anderson T (2013) Development strategy. In: Leach M, Kingsbury D (eds) The politics of Timor-Leste: Democratic consolidation after intervention. Cornell UP, Ithaca, NY, pp 216–238

Bacon R, Tordo S (2006) Experiences with oil funds: Institutional and financial aspects. ESMAP Report No. 321/06. World Bank Group, Washington, DC

Barrowman HM, Kumar M (2018) Conceptions of vulnerability in adaptation projects: A critical examination of the role of development aid agencies in Timor-Leste. Regional Environmental Change 18(8):2355–2367

Courvisanos J (2012) Cycles, crises and innovation: Path to sustainable development—A Kaleckian-Schumpeterian synthesis. Edward Elgar, Cheltenham and Northampton

Courvisanos J, Boavida M (2018) Review of the Roadmap for implementing the SDGs in Timor-Leste: Achievements and limitations. In: Job P, da Silva AB, Mendes NC, da Costa Ximenes A, Soares MB, Niner S, Tam T (eds) New research on Timor-Leste 2017, Volume I. Swinburne Press, Hawthorn, pp 184–193

Courvisanos J, Jain A (2018) Report on commitment of stakeholders to the SDGs in Timor-Leste: Report to the UNDP. Federation University Australia, Ballarat

Dal Poz A (2018) "Buying peace" in Timor-Leste: Another UN-success story? Peace Human Rights Governance 2(2):185–219

Davidson H (2019) Australia accused of Siphoning Millions in Timor Leste Oil Revenue. The Guardian Australian Edition, 16 Apr.

De Haan J (2017) Timor-Leste: An uncertain future and an unsettled boundary. Future Directions International, Dalkeith, WA

Doraisami A (2009) Fiscal policy challenges in Timor Leste: Is the resource curse on the horizon? ASEAN Economic Bulletin 26(2):164–173

Doraisami A (2018) The Timor-Leste Petroleum Fund, veterans and white elephants: Fostering intergenerational equity? Resources Policy 58:250–256

Drysdale J (2007) Sustainable development or resource cursed? An exploration of Timor-Leste's institutional choices. Dissertation, Australian National University, Canberra

Drysdale J (2008) Five principles for the management of natural resource revenue: The case of Timor-Leste petroleum revenue. Journal of Energy & Natural Resources Law 26(1):151–174

Drysdale J (2012) Five principles for managing Timor-Leste's natural resource revenue wisely. In: Langton M, Longbottom J (eds) Community futures, legal architecture: Foundations for indigenous peoples in the global mining boom. Routledge, London, pp 147–160

Hamadi S, Gomes R (2013) Developing Timor-Leste's non-oil economy: Challenges and prospects. Journal of Southeast Asian Economies 30(3):309–321

Hendrapati M (2018) Maritime expansion and delimitation after the Timor Gap Treaty. Indonesia Law Review 5(1):69–87.

IMF (2005) Democratic Republic of Timor-Leste: Article IV consultation. IMF Staff Report, Country Report No. 05/250. International Monetary Fund, Washington, DC

IMF (2009) Democratic Republic of Timor-Leste: Selected issues. IMF Staff Report, Country Report 09/321. International Monetary Fund, Washington, DC

IMF (2012) Democratic Republic of Timor-Leste: Article IV consultation. IMF Staff Report, Country Report No. 12/24. International Monetary Fund, Washington, DC

IMF (2013) Democratic Republic of Timor-Leste: Selected issues. IMF Staff Report, Country Report No. 13/338. International Monetary Fund, Washington, DC

IMF (2016) Democratic Republic of Timor-Leste: Article IV consultation. IMF Staff Report. Country Report No.16/183. International Monetary Fund, Washington, DC

Jensen E (2010) Introduction to the laws of Timor-Leste: The Petroleum Fund Law. Asia Foundation and Stanford Law School, Dili, Timor-Leste

Kent L, Kinsella N (2015) A luta kontinua (The struggle continues): The marginalization of East Timorese women in the Veterans' Valorization Scheme. International Feminist Journal of Politics, 17(3):473–494

Kingsbury D (2012) Challenges of constructing postcolonial unity: Timor-Leste as a case study. Asian Politics & Policy, 4(1):15–32

La'o Hamutuk (2009) Submission to Committee C National Parliament regarding the proposed RDTL General State Budget for 2010. https://www.laohamutuk.org/econ/OGE10/sub/LHSubPNComCOGE10En.pdf. Accessed 16 Apr 2020

La'o Hamutuk (2015) Special economic zone in Oecussi. Resource document. Timor-Leste Institute for Development Monitoring and Analysis. http://www.laohamutuk.org/econ/Oecussi/ZEESMIndex.htm. Accessed 6 June 2019

La'o Hamutuk (2017) Timor-Leste is borrowing more: Will we be able to repay? Resource document. Timor-Leste Institute for Development Monitoring and Analysis. http://laohamutuk.blogspot.com.au/2017/11/tl-is-borrowing-more-will-we-be-able-to.html. Accessed 28 Apr 2019

La'o Hamutuk (2018) General state budget. Resource document. Timor-Leste Institute for Development Monitoring and Analysis. http://www.laohamutuk.org/econ/OGE17/16OGE17.htm . Accessed 16 Nov 2019

La'o Hamutuk (2019a) Timor-Leste buys into the Sunrise oil and gas project. Resource document. Timor-Leste Institute for Development Monitoring and Analysis. https://www.laohamutuk.org/Oil/Sunrise/18SunriseBuyout.htm. Accessed 2 Dec 2019

La'o Hamutuk (2019b) President of the republic vetos state budget. Resource document. Timor-Leste Institute for Development Monitoring and Analysis. https://www.laohamutuk.org/econ/OGE19/veto/PRPRVetoOJE23Jan2019en.pdf. Accessed 2 Dec 2019

La'o Hamutuk (2020) Global economic developments hit Timor-Leste hard. Resource document. Timor-Leste Institute for Development Monitoring and Analysis. http://laohamutuk.blogspot.com/2020/03/global-economic-developments-hit-timor.html. Accessed 14 Apr 2020

Leach M (2015) Politics of history in Timor-Leste. In: Ingram S, Kent L, McWilliam A (eds) A new era? Timor-Leste after the UN. ANU Press, Canberra, pp 41–58

Leach M (2017) Nation-building and national identity in Timor-Leste. Routledge, Milton Park

Lundhal M, Sjoholm F (2009) Population growth and job creation in Timor-Leste. Journal of Asia Pacific Economies 14(1):90–104

Lundhal M, Sjoholm F (2013) Improving the lot of the farmer: Development challenges in Timor-Leste during the second decade of independence. Asian Economic Papers 12(2):71–96

McKechnie A (2013) Managing natural resource revenues: The Timor-Leste Petroleum Fund. Overseas Development Institute, London

Ministry of Finance (2008) Budget review July 2008. https://www.mof.gov.tl/category/documents-and-forms/budget-documents/budget-previous/?lang=en. Accessed 2 Dec 2019

Ministry of Finance (2011) Petroleum Fund annual report fiscal year 2010. from http://www.mof.gov.tl/wp-ontent/uploads/2011/10/PetroleumFundAnnualReportforFinancialYear2010En.pdf. Accessed 12 Nov 2018

NRGI (2013) Timor-Leste: Petroleum Fund of Timor-Leste. Resource document. Natural Resource Governance Institute. https://resourcegovernance.org/sites/default/files/NRF_Timor-Leste_Aug2013.pdf. Accessed 27 Oct 2017

Office of the Prime Minister (2004) Media release: Timor-Leste government launches public consultation on Petroleum Fund. Democratic Republic of Timor-Leste, Dili

Pereira A (2015) The challenges of nation-state building. In: Ingram S, Kent L, McWilliam A (eds) A new era? Timor-Leste after the UN. ANU Press, Canberra, 17–29

Porter D, Rab H (2010) Timor-Leste's recovery from the 2006 crisis: some lessons (Background Note). World Development Report 2011. World Bank Group, Washington, DC

Raimundos O (2019) Oil plan stalled as East Timor President vetoes state budget. AP News, 23 Jan. https://apnews.com/f12d9e12c07e4c2685a52ead a29ace95. Accessed 21 Apr 2020

Scambary J (2015) In search of white elephants: The political economy of resource income expenditure in East Timor. Critical Asian Studies 47(2):283–308

Scambary J (2017) The road to nowhere: The rise of a neo-patrimonialist state in East Timor. In: de la Rama M, Rowley C (eds) The changing face of corruption in the Asia Pacific: Current perspectives and future challenges. Elsevier, Radarweg, pp 267–280

Scheiner C (2015) Can the Petroleum Fund exorcise the resource curse from Timor-Leste? In: Ingram S, Kent L, McWilliam A (eds) A new era? Timor-Leste after the UN. ANU Press, Canberra, pp 73–101

Scheiner C (2020a) Personal email communication based on La'o Hamutuk website information, 13 Feb

Scheiner C (2020b) Implications of recent changes to Timor-Leste's Petroleum Fund, February. Resource documents. Timor-Leste Institute for Development Monitoring and Analysis. http://www.laohamutuk.org/misc/TLS A2019/ChangesPetrolFund.pdf. Accessed 17 Apr 2020

Stewart C (2018) Xanana Gusmao's Timor Sea rant is an own goal for his needy nation. The Australian, 10 Mar.

Timor Gap (2015) Southern coast project. http://www.timorgap.com/databa ses/website.nsf/vwAll/SOUTHERN%20COAST%20PROJECT. Accessed 12 June 2017

Timor-Leste Government (2005) Petroleum Fund Law. Law No. 9/2005. http://www.mof.gov.tl/wp-content/uploads/2011/09/Petroleum-Fund-Law-English.pdf. Accessed 12 Nov 2019

Timor-Leste Government (2009) Petroleum Fund of Timor-Leste Operational Management Agreement. http://www.mof.gov.tl/wp-content/uploads/2011/09/Management-Agreement-25-June-2009-English.pdf. Accessed 12 Nov 2018

Timor-Leste Government (2010) Annex 1, Qualifying Instruments Benchmark & Investment Mandate, Amendment No. 2 to the management agreement between DRTL and Banco Central de Timor-Leste. http://www.mof.gov.tl/wp-ontent/uploads/2011/09/Amendment-to-Management-Agreement-08-October-2010-English.pdf. Accessed 12 Nov 2018

Timor-Leste Government (2011) 2011–2030 Strategic Development Plan. Government of Timor-Leste, Dili

Truman E (2008) A blueprint for sovereign wealth fund best practices/Policy Brief No. PB08-3. Peterson Institute for International Economics, Washington, DC

Wallis J (2014) Constitution making during state building. Cambridge UP, Cambridge
World Bank (2008) Timor-Leste: Poverty in a young nation. World Bank Group, Washington, DC
World Bank (2013) Timor-Leste social assistance public expenditure and program performance report. World Bank Group, Washington, DC
World Bank (2014) Sub-national spending in Timor-Leste: Lessons from experience. World Bank Group, Washington, DC
World Bank (2019) Timor-Leste economic report: Moving beyond uncertainty. World Bank Group, Washington, DC

Practices and Models: Prospects for Commodity-Based SWFs

Xu Yi-chong

Introduction

Since 2000, many resource-rich countries have rushed to set up investment funds to store a portion of their revenues from oil, gas, or other mineral production and exports. Over 50 countries have done so in the past two decades for different reasons—to smooth budgetary volatility, diversify the economy, save for future generations, earmark financial earnings for education, old-age pensions for the poor, or infrastructure, or to mitigate the effects of domestic corruption. This practice of setting up natural resource funds (NRFs) is not without risks, such as diminished incentives to improve expenditure efficiency, reduced budget flexibility, reinforced procyclical fiscal policy (spending goes up in booms and spending goes down in recessions) (Rogoff 1990; Erbil 2011; Bova et al. 2014), or greater complexity of general fiscal management. Indeed, "overenthusiasm and bad advice can leave countries poorer" (Bauer and Mihalyi 2018, p. 1) Whether a fashion or a fad, creating a natural resource fund is a government's policy choice, not a requirement for managing the economy. Thus, some countries rich in natural resources, such as

Xu YC (✉)
Griffith University, Nathan, QLD, Australia
e-mail: yi-chong.xu@griffith.edu.au

E. Okpanachi and R. Chowdhari Tremblay (eds.), *The Political Economy of Natural Resource Funds*, International Political Economy Series,
https://doi.org/10.1007/978-3-030-78251-1_11

Indonesia, did not create resource funds but managed to achieve development; others, such as Mongolia, rushed into creating them but with little long-term success. Mongolia, for instance, established its Fiscal Stabilization Fund in 2011 when revenues poured in from coal, copper, and gold mining; by 2016, however, its economy was on its knees after the real exchange rate depreciated 25% and the fiscal deficit ballooned to 15% of GDP (IMF 2019).

This collection of 10 natural resource funds tells diverse stories of the politics behind the creation, structure, and management of the funds. Some NRFs have existed for decades in their current or other forms, whereas others are relatively new. They are from the world's very rich and the very poor countries, and from democratic and autocratic regimes. Some have been successful in expanding national wealth, helping smooth out national business cycles, and supplanting national incomes, while others are less so (Fasano 2000; Davis et al. 2001). A few simply squandered national wealth as their governments decided to use them irresponsibly. There is no typical fund, as each is the product of the history and politics of the country, and each reflects how the core questions of politics—who gets what, when, and how—were fought and won. Just as there are different kinds of NRFs, there is no single framework that can characterize or even govern them, nor is there one formula with which to preempt potential problems.

All these economies are highly dependent on the production and export of natural resources for national incomes or/and for their foreign exchange earnings. Abundant natural resources have been identified by scholars and practitioners as potentially a mixed blessing for the country concerned. They can provide opportunities for development; but they may also pose serious policy challenges for government—the so-called "paradox of plenty" (Gelb et al. 1988, 2014; Katz et al. 2004; Smith 2004). Some common policy challenges generated by the abundance of natural resources include pressures from the appreciation of the real exchange rate on domestic absorption capacity, rising demands on public spending, and/or increased corruption. Rich natural resources do not guarantee economic successes, nor do they inevitably lead to deteriorating economic performance and development. The political temptation to increase government spending by the whole amount of the extra revenue from resource production and export received each year is often too great to resist which exacerbates these problems.

There are important measures that can be taken to guard against the onset of such problems or to start to reverse them once they have occurred (Stevens 2006). If governments could save some of the revenue and sterilize it by investing it abroad, exchange rate pressures would be reduced, and absorption capacity would not be severely tested. Governments in resource-rich countries can use their resource revenues to stabilize their economies and their budget situation: when oil prices are high, revenues are set aside; when prices fall, governments use the funds to cushion external shocks. To save part of the proceeds from resources offshore can also help "transform finite assets and income from depleting natural resources into permanent wealth in the form of a portfolio of financial assets and its investment income" (Alsweilem 2015, p. 1).

Based on detailed analyses of the nine different NFRs in this collection and with reference to other NRFs, this chapter seeks to draw some comparative lessons on how the government justified to the public assigning a portion of revenue from producing and exporting natural sources to a separate account as an investment fund, the optimal governance structures to ensure the wealth is managed for the interest of the people rather than special interests, in the country, and investment strategies for the funds' sustainability and expansion. One of the important cautionary tales offered in this chapter is that, as there is no typical NRF, there is no one model for all NRFs. There are, nonetheless, lessons countries can draw on whether and how to set up NRFs and how they are expected to operate. A clearly defined legal framework, transparent governance structure, and rules-based investments are all helpful, but they do not determine the success or failure of NRFs. Understanding the forces behind each stage of the fund's development is what this chapter seeks to achieve.

Creating a Natural Resource Fund

Natural resource funds can be established by one of three routes:

- A constitutional amendment: so far, the Alaska Permanent Fund is the only one to have been set up through an amendment to the state constitution, because an existing clause in the state constitution of Alaska banned any expenditure earmarking of natural resource revenues. This typically requires the greatest amount of political support and would subsequently require similar support to undo any such legislation. In other words, there would have to be a political commitment not only of the current government but the people in general.

- An act of Parliament: several countries have established natural resource funds by an act of Parliament (e.g., Alberta [Canada], Botswana, Chile, Ghana, Norway, Timor-Leste). In order to do so, Parliament usually provides an opportunity for public discussion of the aims of the fund and the broad outline of its operation. The legislation enacted often delegates details of the management of the fund to an executive agency such as the ministry of finance. This arrangement combines a degree of stability with the possibility of flexibility. Major changes to the act on issues such as the objective of the fund or even its termination will require subsequent legislation, while other desired changes can be made with the agreement of the executive agency and the management of the fund without recourse to Parliament. Alberta fits in this category.
- An executive decree: the most expeditious way to establish a resource fund is by decree, as was done in Azerbaijan Saudi Arabia and Venezuela, which in most cases does not require the approval of any other government body. By the same token, decrees can be used readily and quickly to reverse or alter previous decisions, as seen in Venezuela.

Several caveats about this categorization of the establishment of NRFs need to be noted: One is that no matter what method is used to create a fund, by definition, an NRF can only be established by government actions. Unlike other financial institutions, NRFs are owned by the government on behalf of its citizens: the government has an exclusive property right over an NRF, and an exclusive right to decide its governing structure and legal and political accountability (Hart 1995; La Porta et al. 1998) and define its mandates, obligations, and rent-seeking opportunities (Williamson 2000). Second, in some political systems, the line between a Parliament act and an executive decree is nonexistent. This is the case in many Gulf states, whether they are federations of monarchies as the United Arab Emirates or an absolute monarchy as Saudi Arabia (Roll 2019). Finally, even in democratic political systems, the balance between the legislature and the executive varies significantly, due in part to the original allocation of power in the constitution and in part to the historical evolution of the political system. The methods by which NRFs are created need to be taken into consideration but they can only partially explain the existence and operation of an NRF.

The legal status given to an NRF is an important factor to explain its chain of accountability and responsibility. Some NRFs were set up with independent legal status and are registered under commercial law as corporations or firms. In such cases, the government transforms its ownership role into that of an overseer (Gelpern 2011; Clark et al. 2013). Questions as to whether a separate legal status makes any difference in an NRF's operation and performance are highly contested. For instance, the Kuwait Investment Authority (KIA), established before the country even become independent in 1953, does not have an independent legal status, but has always insisted that it operates without political interference. Although both the Government Investment Corporation in Singapore and the China Investment Corporation were established as incorporated firms under their company laws, few consider them being independent in their operation (Balding 2012). In contrast, neither the Government Pension Fund Global (GPFG) in Norway, the Pula Fund in Botswana, nor the ESSF in Chile has a separate legal status; yet they are considered to operate without direct political intervention even though they are accountable to the Minister of Finance, who is in turn accountable to the legislature.

Moreover, in setting up NRFs, there were extensive public debates and public campaigns in some places, while in other places, the government took the action as part of the macroeconomic or fiscal policymaking without additional public debates. In Alaska, for instance, the initial oil revenues were used primarily for general economic development and economy diversification for over a decade, but the money was quickly spent with little apparent success. Public campaigns were organized in the early 1970s, demanding oil revenues be turned into a long-term income stream by creating a permanent fund. The demand eventually won the support of two-thirds of the population (Smith 1991; Hannesson 2001). In Alberta, similar debates over oil revenues took place too but the demand to create a resource fund was driven by different forces and for different reasons. The debate reflected a long-held sense of "western alienation" and the rise of "Prairie Capitalism" while Alberta's petrochemical companies were becoming multinational forces (Pratt and Richards 1979; Stevenson 1981). The initial debates or their absence in the process of setting up NRFs can partially explain their later development.

In Timor-Leste, the public debate started *before* any oil revenue came in. Timor-Leste became an independent state in May 2002, and it was among the poorest countries in the world. Its GDP per capita was less than $300 in 2002, but its population of one million sat on rich oil reserves in the Timor Sea between Timor-Leste and Australia. The controversial negotiations, marred by allegations that "Australia, one of the largest and most powerful countries in the Asia–Pacific region, recently legalized its thievery of tens of billions of dollars in resources from one of the smallest and weakest" (Scheiner 2006, p. 30; Scheiner 2005) led to an agreement of a 90-10 split between Timor-Leste and Australia over the revenue from one of the three oil fields in the region (IMF 2004, p. 5; Hood 2005). This immediately raised the question of how to manage the sudden inflow of revenues in Dili. While there were pressing needs for government spending on infrastructure, education, health, poverty alleviation, and other concerns, the economy had a very limited capacity to absorb a large inflow of revenues and manage future demands. The government put the issue to the public for consultation. After receiving extensive inputs from the public, multilateral institutions, aid agencies, and NGOs, the Parliament approved the Timor-Leste Petroleum Fund Act in July 2005, and the Timor-Leste Petroleum Fund went into operation that September. The extensive public consultation led to the adoption of the Act with its main features based on the "Norway plus model": clearly stated objectives, governing structures, and operational principles. As in Timor-Leste, government agencies in Ghana started debating what to do with the oil revenue *before* oil production started. This practice of preparing to set up a fund before production starts was strongly supported and encouraged by the IMF and the World Bank.

By contrast, in Chile and Norway, this process was long and drawn out. In Norway, the establishment of an investment fund came nearly two decades after the initial discussion of North Sea oil as a potential source of revenue. In Chile, there were no public debates or campaigns about whether to park some of the resource revenues into an investment fund, but the question of how to manage volatile national income from copper production and exports had been a core issue for the government long before the Copper Stabilization Fund was established in 1985, a predecessor of the Social and Economic Stabilization Fund (ESSF). Thus, the absence of public debate in Chile and Norway is not a sign of a lack of public participation: the public was part and parcel of the general and continuing processes of macroeconomic and/or fiscal policymaking.

Just as the Fiscal Responsibility Law of 2006 institutionalized Chile's ESSF and Pension Reserve Fund, the Fiscal Responsibility Law of 2007 in Nigeria authorized the national government to create a legal framework for centralized savings of oil revenues in the Excess Crude Oil Account (ECA). Unlike in Chile, where the Law was implemented without controversy, in Nigeria, the authority of the national government to set up an investment fund with national resources and the legality of the ECA were both challenged on the ground that the ECA was illegal because it was not "backed by law on the verdict of the Supreme Court." The debate continued after the Nigeria Sovereign Investment Authority (NSIA) was set up with the transfer of assets from the ECA. The real issue, however, was not the legality of the ECA or Nigerian Sovereign Investment Authority, but the competing demands of oil revenues between the national and state governments and among state governments. Fiscal federalism over the years was behind intense debates, constant struggles, and even violence in the form of agitation for resource control in the oil-bearing Niger Delta Region (Amusan et al. 2017, p. 453; cf. Arowolo 2011; Salami 2011).

In Saudi Arabia, despite broad recognition that the economy in Saudi Arabia must change, it took a dramatic collapse of global oil prices in 2014 and an emergence of a young leader to make a significant policy shift regarding its oil resources and oil revenues. On 25 April 2016, then-Deputy Crown Prince Muhammad Bin Salman (now the crown prince) revealed an audacious economic and social reform plan. The ambitious Saudi Vision 2030 established a broad strategy and goals for the proposed reform that was designed to build not just a thriving economy, but also a more vibrant Saudi society and a higher-performing, modernized government. To diversify its oil-dependent economy, the Vision 2030 highlighted the need to transfer under-utilized state assets of past and current oil revenues to aggressive state-owned investment funds to finance the reform and to create a future stream of public revenue in order to support living standards independent of oil export revenues (Alsweilem 2015; Roll 2019; Grand and Wolff 2020). The Kingdom had set up the Saudi Arabian Monetary Authority (SAMA) back in 1953. SAMA, like other foreign exchange authorities, such as that in Hong Kong, was not designed as an active investor in the global financial market. This made Saudi Arabi different from its neighboring countries, Kuwait, Bahrain,

Qatar, United Arab Emirates, and Oman, which all have active NRFs. With Vision 2030, the kingdom would transfer assets from SAMA to PIF to maximize "returns on the kingdom's assets" and to use returns on PIF investments available to support public finances, ultimately independent from its oil production and export.[1]

Sudden changes of this nature are always controversial.[2] It is not surprising that the Saudi Vision 2030 was discussed and scrutinized much more outside of Saudi Arabia than inside the kingdom. The importance of public debates prior to the establishment of NRFs is that they can reveal much of the political issues behind NRFs.

NRFs can be set up to achieve several objectives: to save, to stabilize, or to expand and develop:

1. To save—saving revenues is an objective because most natural resources are exhaustible. Several early NRFs clearly show the importance of saving while resources last. In 1956, the British colonial administration of the Gilbert (now Kiribati) and Ellice (now Tuvalu) Islands set up a Revenue Equalization Reserve Fund in Kiribati in anticipation of the expected exhaustion of phosphate minerals, which occurred in 1979. Some other Pacific islands, such as Nauru, were rich in phosphate, and the mining of phosphate for fertilizer supported successful agriculture industries in Australia and New Zealand in the 1950s and 1960s. However, with 90% of the islands' surface stripped away between 1900 and 1979, those islands were permanently ruined, and the people were forced to relocate to other

[1] For instance, the central bank in Saudi Arabia in 2020 at the peak of COVID-19 and amid the decline of oil price made a further transfer from SAMA to PIF of $15 billion in March and $25 billion in April (see Rashad 2020).

[2] For instance, at the time the Canadian government was contemplating moving part of the Canadian Pension Fund into an active investment fund, similar proposals were made for the Social Security Fund in the USA. The former became a political reality in 1997 when the federal government made its initial transfer of $12.1 million to the fund. By 2001 when it went into operation, several more transfers were made and the assets under its management reached $100 billion in 2006 and $395 billion by the end of 2019, supporting the future liability of old age pension and disability payment. In contrast, the opposition in the USA prevented the social security trust funds from becoming active investment funds on the grounds of private responsibility. The size of its assets under its management remains the largest in the world, at $2890 billion in 2020; its liability is the largest as well. According to the Congressional Budget Office, the funds would start to level off by 2016 and run out by 2030.

places, including Fiji and New Zealand, and Nauru and several Pacific Islands had to fight in court for compensation in order to support their remaining populations (Burnett and Lee 1980; Edwards 2014; Gale 2016).

2. To stabilize—some NRFs were established with the explicit objective of stabilizing the business cycle or/and boom-bust cycles of resources. In Chile, the Copper Stabilization Fund was created in 1985 to help government stabilize the budget revenue in light of the high volatility of world copper prices. One key question about a Stabilization Fund is its sustainability, especially when the resource price is on a downward trend. In 2017, for instance, "Kazakhstan had a gross government-debt-to-GDP ratio of around 20% and maintained a strong overall fiscal position" through its NFR—that is, a transfer from the NRF to state coffers to main a fiscal balance (EBRD 2018, pp. 15–16). Such a model may not be sustainable because of a large scale of fiscal imbalance, potential depletion of resources, and/or sudden price falls. Kazakhstan has already witnessed the rise and rapid fall of the assets of its NRF. To make a Stabilization Fund work and be sustainable, there has to be a combination of objectives of stabilization and saving.
3. To expand—some funds, such as the Alaska Permanent Fund and the Public Investment Authority (PIF) in Saudi Arabia, were created to grow the wealth of the country, an objective of many sovereign wealth funds of other asset sources.

These objectives are not mutually exclusive. Indeed, funds are often created to achieve a combination of objectives, such as saving and stabilizing, saving and expanding, or saving and developing. The usefulness of this categorization of the objectives of NRFs is apparent because they are closely related to investment strategies: that is, stabilization funds tend to have more conservative investment strategies, as they must be liquid so that government can use them in time of need. As stabilization funds, NRFs must be integrated into the national fiscal policy and budgetary process, which sometimes presents challenges for those who manage the fund.

Governing Structures of NRFs

The creation of an NRF often highlights the link between resource production and resource utilization, making the fund a useful political and policy tool. Its governing structure is critical to guarding against squandering national wealth, either due to incompetent management or to political corruption. Governance, transparency, and accountability of NRFs and sovereign wealth funds (SWFs) in general are as, if not more, important to countries that are to host their investments (Sauvant et al. 2012). The common belief is that if proper rules and governance structures are put in place, NRFs can work for the benefit of all. That is, an NRF needs a legal framework with clearly delineated rules about its objectives, deposits, and withdrawals, and accountabilities so that the fund can operate as an investor in domestic or overseas financial markets.

Most SWFs whose initial capital came from government budget surpluses, such as the Future Fund in Australia, or public pension funds, as in Canada and New Zealand, or foreign exchange reserves, as in China, Singapore, and South Korea, were often set up with independent legal status, registered under their respective countries' company law or commercial law. These funds may manage national wealth, but they do so with the government overseeing their operation through its representation at the board, rather than direct involvement in daily decisions. A few NRFs, such as the Alaska Permanent Fund and the Alberta Heritage Savings Trust Fund, are organized in a similar way. Most NRFs, however, are set up at a unit under the ministry of finance or the central bank. This is in part because many serve the function of assisting in the stabilization of government budget cycles. Their legal status may not be as important as the rules governing the funds, which define the organizational structure of the fund and its accountabilities, outline payments into and out of the fund, and determine the rules in managing the fund.

The governing structure of NRFs consists of both horizontal and vertical accountability. Vertical accountability is demonstrated in its reporting line of the management of the fund, which continues up the hierarchy until it meets a minister. Parliament holds the responsibility of overseeing the fund. There is a large degree of difference in terms of parliamentary involvement; for instance, in Norway, Parliament plays an active role in not only overseeing the fund but setting the guidelines for its operation and investment, especially through its ethics committee (Moses and Letnes 2017; Lie 2018). In Chile, Parliament is involved in the budget-making process. In most other countries, democratic or not, legislative oversight is infrequent and superficial, in part because managing NRFs has already been delegated to the executive and in part because of the lack of expertise and competence in NRFs among parliamentarians. Horizontal accountability operates through two mechanisms: elected or appointed officials independent of government receive regular reports on portfolio performance, and readily available information published in annual reports, press releases, and audit reports (see Petersen and Budina 2003).

Transparency and accountability are expected from SWFs, especially after the international working group of SWFs adopted the Generally Accepted Principles and Practices (GAPP) of SWFs, known as the Santiago Principles, in 2009. Despite initial resistance to negotiating rules, nearly all SWFs/NRFs have signed on the GAPP. This is especially important for relatively new NRFs in developing countries, which require credibility in order to invest in good assets elsewhere.[3] The international working group of SWFs was institutionalized into the International Forum of SWFs (IFSWF). Some initial participants of the negotiation in Santiago, notably Norway and Chile, withdrew from IFSWF, but others from new and developing countries rushed to join once they became operational. IFSWF consists of members of SWFs/NRFs from developed economies, such as Australia, New Zealand, Ireland, France, and Italy; Gulf oil states, such as Kuwait, Iran, and Qatar; many countries in Africa, such as Angola, Ghana, Nigeria, and Senegal; and Pacific and Caribbean island states, such as Nauru and Trinidad and Tobago. All members of IFSWF have made formal commitments to implementing GAPP and conducting annual or biannual self-assessments of their compliance to the Santiago Principles. NRFs are also expected to make efforts to improve their transparency, accountability and rule-based performance and investments. A variety of organizations nowadays scrutinize the operation of the management of NRFs, but there is a huge gap in transparency ranking among NRFs.[4]

Since NRFs are investment funds, clearly delineated rules on payments into and out of the funds are essential. The size and growth of NRFs

[3] For more on the negotiations on the Santiago Principles, see Truman (2010).

[4] The SWF Lab at Universita Bocconi in Italy and SovereigNet at Tufts University and IE Foundation conduct biannual analyses on these funds, as do several consulting firms such as the Boston Group, PWC, and Rhodium Group.

are determined either by overall fiscal policy or by the rules adopted when the fund was set up. Often the initial lump-sum instalment made by the government into the fund would be followed by annual transfers from resource revenues. The fiscal rules in Chile and Norway, for instance, decide that when resource revenues collected by the government are higher than the non-resource government deficit, the surplus would be deposited into the investment fund. When resource revenues fall and the government budget deficit rises, the government can withdraw from the fund to supplement spending. Rules regulating payments into and out of NRFs vary across countries, but they need to be clearly set and transparent to ensure public confidence that the national wealth will not be squandered and the fund can be trusted as legitimate investors in global financial markets. In some resource-rich countries, an independent body of experts advises the Finance Minister on the future price of resources based on which government can plan for its budget. In so doing, the government can strengthen its fiscal rules by not politicizing the calculation of the structural balance of future incomes and expected expenditures.

Putting a certain amount of national wealth into an investment fund is controversial, while rules on withdrawing from it are critical for the funds' survival. One important concern is that a lack of clearly defined and implementable rules could lead to depletion of the fund or misuse of the fund by political elite as seen in Angola. There are several types of withdrawal rules in place: a given proportion of resource revenue may be paid into the fund and all income and capital must remain in the fun and may not be withdrawn until legislation provides otherwise; some or all of the income of the fund is allowed to be used but not the capital; and both income and capital can be withdrawn. How much, when, and how resources from NRFs can be withdrawn are normally specified by the country's fiscal rules, which require legislative approval. These rules vary greatly across countries (see Bauer 2014). For those NRFs that are set up for stabilization objectives, withdrawing resources from the fund is often part of the budget-making process and thus depends on the politics of the country and of the day. Therefore, not only should "the rules of transferring funds between an SWF and its owner … always be spelled out" (Al-Hassan et al. 2013, p. 8), but there must also be strong political discipline to delink the spending from the resource revenue. For those permanent funds, as in the cases of Alaska and the PIF in Saudi Arabia, withdrawal depends on the returns of fund investments. In both cases, it is important to have these rules clearly stipulated before the fund is in operation. The key is not whether a proportion of the NRF could be used or not for domestic purposes. Several NRFs in this collection allow the government to withdraw some resources to make up the budget shortfalls (e.g., Chile) or to finance domestic infrastructure projects (e.g., Ghana). In these cases, rules were put in place on when, how and under conditions withdrawal of part of the resources could take place and rules could be implemented. What allowed the President of Angola and his relatives to have misused the fund for personal use was not rules, but the entire system that allows a few political elite to conducted what was known as a master class in corruption along the natural resource value chain (OECD 2016).

NRFs consist of a diverse group of funds that are governed by different rules on sources of funding and on withdrawal. For instance, the Pula Fund in Botswana has two concurrent deposit rules: one determined by the Bank of Botswana and one by the government. In Ghana, different rules on withdrawing resources have been put in place for three different funds: the Petroleum Holding Fund, the Ghana Heritage Fund, and the Ghana Stabilization Fund. Withdrawals cannot be made from the Ghana Heritage Fund, for instance, until the oil reserves are depleted and the Petroleum Fund's assets are merged into the Ghana Petroleum Wealth Fund. In contrast, if quarterly oil revenues fail to cover 25% of the annual budget funding amount, the government can withdraw resources from the Ghana Stabilization Fund (Alsweilem et al. 2015). Rules nonetheless are often up for interpretation. As the chapter on Chile shows, even the definition of a "rainy" reason can be debatable. A similar case can be observed in the history of the GPFG in Norway: did the economic situation in 2008–2009 justify the withdrawal of extra resources from the fund to stimulate the economy? There was no one answer, and both the politicians and the public debated this question (Tranøy 2010). Moreover, a much more subtle issue often emerges in the debate on withdrawing resources from NRFs: those who manage the fund tend to resist any withdrawals. As Peter Castello, the chairman of the Board of Guardian of the Future Fund in Australia, warned the government during the global financial crisis: "Don't ever think about touching it." They often become defenders of the fund for the sake of defending it (see Xu 2017).

Another important question is whether the government can, or should, use an NRF as collateral in its borrowing, especially from international markets. Although many NRFs prohibit borrowing against the fund's resources, the government may be able to increase its debts by using part of the asset of NRFs as collateral. Initially, the Timor-Leste Petroleum Fund was prohibited from being used as a guarantee on public debt. In 2013, the rules changed, so that 10% of the Fund might be used as a collateral. This reduced interest rates on loans while still protected 90% of the fund from any potential consequences of poor public financial management (Rietveld and Bauer 2014, p. 10). Discussions on the topic also took place in Ghana. When the government of Chile could not withdraw resources from the ESSF to meet rising demands in expenditure, especially in supporting its state-owned copper company, it borrowed from international markets with the support of Parliament. The warning in these cases is that, where the cost of the debt is above the return on government investment, this option should be avoided. This is a delicate and often politically controversial balance. Several countries borrowed from the IMF during the COVID-19 pandemic rather than using the sources of their NRFs, either because of the restrictions on withdrawal or for other reasons.

In sum, the governance structure of NRFs is critical in making them sustainable and in preventing rentier behavior by the political elite. Rules on payments and withdrawals, especially for budget stabilization purposes, can be formulated, but, more importantly, they need to be adhered to in times of need. A fund whose original purpose was to provide some resources for future generations could be easily extinguished if all resources were withdrawn without legislative change. Strong political discipline and adherence to long-term objectives are required for NRFs to remain sustainable.

Allocating Strategic Assets of NRFs

Once NRFs are established, the management of these assets becomes an important issue. The strategic asset allocation is related to the policy objective and its associated risk constraints. There are four categories of investments that NRFs can make:

- Cash: highly liquid and low risk, low returns from money market instruments, such as short-term government bonds and bank deposits
- Fixed income: other debt instruments with slightly more risk and return, such as government or corporate bonds

- Equities: stocks in companies with varying degrees of risks and returns
- Alternative assets: more volatile and complex assets with higher long-term expected returns, such as real estate, infrastructure, derivatives, and private equity.

Stabilization funds tend to allocate more assets in cash and fixed income, while saving funds can afford longer-term, higher-risk investment in alternative assets. In general, the higher-risk assets require better capacity to manage its assets. Several important and practical challenges facing all NRFs are who manages the strategic asset allocation; where can and do NRFs invest; and how NRFs maximize returns while being held accountable. These are complex issues as NRFs have become mature investors in global financial markets, which are primarily dominated by competition among private investors. As discussed above, all NRFs have a similar governing structure that combines simultaneous horizontal and vertical accountability. The governing board, which may be known by other names, such as "council" or "board of guardians," sets the guidelines for strategic asset allocation; the fund management team either makes allocation in-house or selects third-party fund managers to do so. Some funds, such as the ESSF in Chile, the GPFG in Norway, and the Timor-Leste Petroleum Fund, are prohibited from investing in domestic markets, where others can invest a portion of their assets domestically. Investing resource revenues in global financial markets reduces the pressure on their currency and domestic capacity if they have to absorb a sudden inflow of foreign currency (often in US dollars, which is behind the so-called resource curse).[5] (Table 1).

Most NRFs are more conservative in their investments than sovereign wealth funds of other sources of assets (Table 2).

Changes have taken place in the asset allocations of NRFs. Taking Chile's two NRFs as examples, the PRF and the ESSF were created to facilitate fiscal stability and stabilize the exchange rate in the event of an external or internal shock. To fulfill this objective, these two investment

[5] When a portion of an NRF is invested domestically, there may be a significant difference between returns in local currency and those in USD (e.g., in 2019, ESSF earned a 12.6% in Chilean pesos while it lost a 0.4% in USD) and there is no rule in terms of the specific financial indicator that must be used. Caution is therefore needed in analyzing the performance of NRFs.

Table 1 Asset allocation of selected NRFs, 2017

	Medium-term return target	*Target asset allocation*			*Average annual nominal return, 2010–2017 (%)*
		Money market & fixed income (%)	*Equities (%)*	*Alternative assets (%)*	
Alaska	5% + CPI	22	51	23	9.7
Alberta	4.5% + CPI	20	50	30	11.5
Chile ESSF		92.5	7.5		1.3
Chile, PRF		85	15		3.0
Norway	3% + CPI	35	60	5	7.6
Timor-Leste			90	10	3.5

Source Natural Resource Governance Institute, 'How Good are Sovereign Wealth Funds at Investing Money Made from Natural Resources?' 13 June 2018, at https://resourcegovernance.org/blog/how-good-are-sovereign-wealth-funds-investing-money-made-natural-resources

Table 2 Asset allocation of the Future Fund (FF) of Australia and the Government Investment Corporation (GIC) of Singapore, 2010, 2016

	Equities		*Fixed Income*		*Alternative assets*	
	2010 (%)	*2016 (%)*	*2010 (%)*	*2016 (%)*	*2010 (%)*	*2016 (%)*
FF	35	33	37	29	28	36
GIC	38	31	41	30	29&	39

Source PwC, "The rising attractiveness of alternative asset classes for Sovereign Wealth Funds," 2018

funds have short investment time-horizons and investments that tend to be liquid. Chile's Economic and Social Stabilization Fund (ESSF) is a good example of a stabilization SWF. Founded in 2007 to repay public debt and fund fiscal deficits, it kept its original objective over the years. Until 2013, the SWF could invest exclusively in bonds and money market instruments. Given the decreasing returns delivered by fixed income, the fund's board decided to introduce important changes in the investment policy. From that year on, the SWF was allowed to invest in equities, which now represent 7.5% of the portfolio. The remaining part of the portfolio is composed of sovereign bonds (55%), inflation-linked sovereign bonds (3.5%), and money market instruments (34%).

In recent years, the ESSF has adopted a more aggressive investment strategy to achieve an expected annual return of at least 2% above its inflation rate. It has reduced its holding of sovereign bonds from 66.5% in 2008 to 23% in 2017, while increasing its holding of higher-yield equities, bonds, and real estate assets (Chilean Ministry of Finance 2017) (Fig. 1 and Table 3).

This is the trend of investments by many SWFs/NRFs in the era following the global financial crisis of 2008. Since 2011, SWFs/NRFs have increased their shares in alternative assets. One-third of property markets in the world, including SWFs/NRFs, are owned by institutional investors. The GPFG in Norway, for instance, has been identified as "a perennial investor in global real estate with an original target of 5% of its assets" and 16% of the total transactions made by world SWFs in 2018–2019 were in property markets (Capapé et al. 2019, p. 29). Two

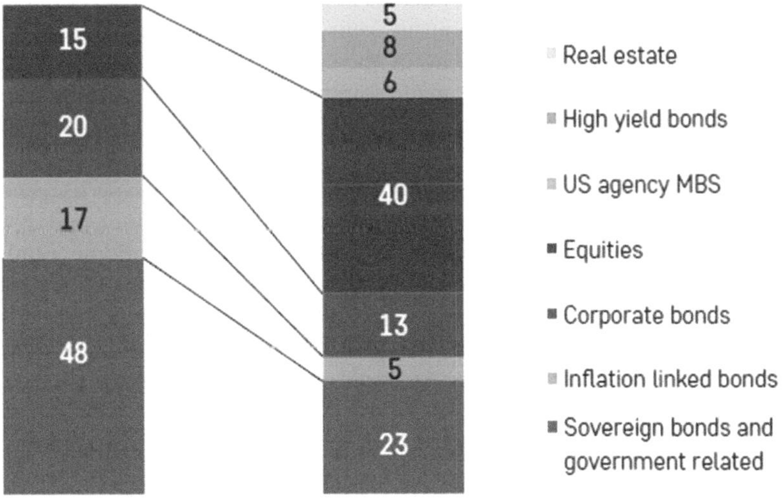

Fig. 1 Asset allocation of ESSF of Chile, 2008, 2017. *Note* MBS—mortgage-backed securities (MBS) issued by US agencies (*Source* Chilean Ministry of Finance [2017])

Table 3 Market value of the PRF and the ESSF (in US$ million): 2007–2017

Year	*PRF*	*ESSF*	*Total market value*
2007	1,466	14,033	15,499
2008	2,507	20,211	22,717
2009	3,412	11,285	14,706
2010	3,837	12,720	16,557
2011	4,406	13,157	17,562
2012	5,883	14,998	20,881
2013	7,335	15,419	22,754
2014	7,944	14,689	22,633
2015	8,112	13,966	22,078
2016	8,862	13,772	22,634
2017	10,000	14,700	24,700

Source Elaborated by the author from data available on the Chilean Ministry of Finance's SWFs annual reports: https://www.hacienda.cl/english/sovereign-wealth-funds/annual-report.html

associated issues need some attention: first, as interest rates have been at record lows, it seems to make sense for many SWFs and NRFs to invest in real estate. When the share of their total investment went from 24% in 2013 to 46% in 2014, property markets in pathway cities such as London, New York, Sydney, and Melbourne went through the roof. There are broader impacts on the economy in these cities and countries. Second, most property investments are made jointly with other SWFs/NRFs, or with other institutional investors, an issue that will be discussed in more detail below.

Investing in global financial markets requires knowledge, expertise, and human capital to ensure the asset value of NRFs. Therefore, initially, nearly all SWFs and NRFs outsource fund management. Some funds continue this practice even when they become mature investors; they have a small team of professionals who are able to identify fund managers around the world to do their strategic asset allocation for them. The Future Fund in Australia is one example of outsourcing, with 100% of its asset allocation delegated to global fund managers. Such an outsource management of asset allocation was built into the initial governance structure in Timor-Leste. The Petroleum Fund of Timor-Leste is managed by three parties:

- The Investment Advisory Board develops overall strategy, investment mandates and benchmarks and risk management;

- The Executive, being the Minister of Finance on behalf of the government, makes all decisions except external manager selection;
- The Central Bank executes the decisions of the executive and makes the external manager selection decision.

The legislation that created the fund appointed JP Morgan Chase Bank to serve as custodian institution for the Petroleum Fund. JP Morgan selects external managers for all its equity investment and 20% of bond investments; the rest is managed by the central bank. Timor-Leste is not unique in contracting out fund management; many other developing countries do the same (Chatham House 2014). For example, the International Trust Fund Agreement, signed between the governments of Tuvalu, the UK, Australia, and New Zealand in 1987 to set up a Tuvalu Trust Fund to provide additional budget support and generate a stream of income for future generations, has been managed by two fund management firms in Australia. NRFs from developed countries may also outsource their fund management: in Norway, the GPFG used to outsource 80% of its assets to professional fund managers but has in more recent years invested heavily in building in-house capacity, so that its in-house team now makes all decisions on asset allocation. This change took time, often two to three decades, and serious investment in human capital.[6] The GPFG did so in part because to control the risk, in part to meet ethical standards set by the Parliament, and in part to internalize the cost.

Outsourcing strategic asset allocation is seen as a positive step to demonstrate that SWFs/NRFs have become mature and as "legitimate" investors in global markets as other institutional investors, such as pension funds, private equities and even venture capital firms (Clark and Monk 2013). Consequently, they should and could be used to help development, long-term infrastructure expansion, and other objectives, such as the aims of the One Planet Sovereign Wealth Funds: integrating climate change risks and investing in the smooth transition to a low-emissions

[6] An interview with the chief executive office of the Government Investment Corporation, Singapore, reveals that it took three decades for the GIC to build its in-house capacity, and that an understanding of the global financial markets was necessary for the GIC to make right choices (see Wright 2019).

economy.[7] When government delegates responsibility for the management of the fund to the central bank that then engages in third-party custodians and specialist asset managers for investments of the assets, it presents some serious challenges of accountability. Any given organization has a set of cascading principal-agent relationships. In "hiring" an agent, a principal can put in place a set of mechanisms to ensure the agent does what the principal wants it to, such as clearly delineated rules for rewards and punishments, monitoring and reporting requirements, screening and selection procedures, institutional checks and sanctions. Meanwhile, at each level, agents operate with significant expertise and access to information. Agent specialization exacerbates the twin problems for the principal of hidden information and hidden action. The longer the principal-agent chain is, the more difficult it becomes to hold players accountable.[8]

Furthermore, asset management mandates may be structured based on global total returns objectives, or on a benchmark representing a specific investment universe. Even when adequate mechanisms are put in place for the management of NRFs to control and monitor the actions of third-party fund managers and all steps—hiring, performance review, subsequent fund allocation decisions—are transparent, the expected objectives of asset managers may not be in line with the ultimate objective of the fund. In other words, fund management firms are expected to maximize returns and that is the reason they are selected. In-house decisions to invest may have to include other considerations, the balance that the government ultimately must make. For instance, in 2018, the return for the investment of GPFG in Norway was −6.1%, while the Future Fund in Australia had an annual rate of return of 5.8%. The Future Fund outsources all its strategic asset allocations and its strategy was justified to Parliament on the ground that its objective was to maximize its returns. When asked whether the Future Fund would consider divesting from the coal industry, the chairman of the board unequivocally answered, "of course not." This was in great contrast to the important divestment

[7] The One Planet Sovereign Wealth Funds is a network of six large SWFs from UAE, Kuwait, New Zealand, Norway, Qatar, and Saudi Arabia to support and finance firms taking on low-carbon projects. See https://oneplanetswfs.org/.

[8] There is no shortage of literature on principal-agent problems by economists, political scientists and other social scientists. See, for instance, Weingast and Moran (1983), McCubbins (1985), Adams (1996), and Miller (2005).

decisions made by the GPFG, such as one from Walmart for labor standards, from VW for its violation of emission standards, and from the coal industry for its role in climate change. These changes were expected by the Norwegian citizens; conversely, most Australians do not even have an idea what the Future Fund is about. It is not that the government tries to hide it, but the Future Fund ensures that it would be run as a financial institution with little public inputs. Outsourcing strategic asset allocation remains an issue to be examined further across NRFs. One piece of evidence is that some large SWFs/NRFs have been building in-house capacities to reduce their dependence on third-party fund managers.

Another emerging issue is co-investment. In the past decade, while most SWFs/NRFs have improved transparency and governance in part to be seen as legitimate investors (Stone and Truman 2016; Truman 2017), many, especially the larger ones, have changed their investment strategies. They have taken on riskier investments in so-called alternative investments, investments in assets that are traded in private, rather than public markets, such as real estate, private equities, or venture capital. They have also increased co-investments with "like-minded" investors, which can be other FWSs/NRFS, pension funds, or even hedge funds (Bernstein et al. 2013; Cumming et al. 2016; Monk et al. 2017).

Since 2014, for instance, SWFs/NRFs have increased their co-investments in venture capitals in bio-health technology start-ups or financial technology start-ups. 2018–2019 was a record period for SWFs/NRFs co-investing in venture capital, pouring in $30 billion and making up 9% of the world's total venture capital investment (Capapé et al. 2019). For instance, Ant Finance, a FinTech subsidiary of Alibaba, received a co-investment of $10 billion in 2018 from the two SWFs in Singapore (the Government Investment Corporation and Temasek), the Canadian Pension Fund, and several US private equity firms such as Warburg Pincus, Silver Lake, and General Atlantic, and others. Some argue that co-investments of NRFs and other financial institutions can help reduce risks and control costs (Johan et al. 2012; Dixon and Monk 2014). Others argue that co-investments offer a large pool of capital that can and should be tapped for "good use," such as climate change and infrastructure projects (Thatcher and Vlandas 2016; Monk et al. 2017; Caldecott and Harnett 2019). Still, others contend that co-investments led by SWFs/NRFs disguise state capitalism, undermine private investments, and threaten the *prima facie* logic of financial capitalism (Wood and Wright 2015; Dolfsma and Grosman 2019). These different views highlight many known unknowns, and the need to appreciate potential developments that may affect us all. Although we know that risks in venture capital investing in start-ups remain high (Zider 1998), we do not know how NRFs would be able to balance long-term and short-term risks with public money. Similarly, although we know that SWFs/NRFs were created by governments and state ownership is therefore their core, we do not know whether and how they can be held accountable by their governments once they engage in co-investments; nor do we know how this new class of state players is reshaping the operation of global financial markets that are dominated by private firms and regulated by private rules.

In the increasingly complex choreography of the world economy, SWFs play an increasingly important role due to their capacity to channel large volumes of savings toward investment. Yet, as SWFs/NRFs have become more adaptive to technology innovation, they have increased investment in sectors such as biotechnology, software, fintech, data, mobility, e-commerce, and health. Large economies drew a lion's share of the investment made by SWFs/NRFs: in 2018–2019, 38% of investments went to the USA, 13% to China, 8% to India 8%, 6% to UK 6% and 4% reach to Singapore, Ireland and Russia (Capapé et al. 2019, p. 8), while little went to fill up the annual investment gap of $2.5 trillion to achieve the Sustainable Development Goals (United Nations 2017). In the time of need, such as during the pandemics in 2020, countries such as Timor-Leste and Ghana again relied on borrowing from the IMF or the World Bank, rather than withdrawing resources from their respective NRFs.[9] The questions of whether and when NRFs should be used for domestic development remain highly contentious issues.

Conclusion

NRFs have become more popular, especially among developing countries, to manage several challenges, whether long-recognized pressures on exchange reserves, volatility of resource prices and external shocks or rent-seeking behavior of political elites. NRFs remain a diverse group of investors, and their performance varies significantly. Their creation is not panacea to the social, economic, and political problems a country faces;

[9] In the first half of 2020, IMF approved emergency loans to Chile (24 billion), Ghana ($1 billion), and Nigeria (3.46 billion).

indeed, NRFs are not necessarily solutions at all. As all resources belong to the people in the land, what to do with revenues from resource production and exports has always been hotly debated at three stages: whether and how to set aside a portion of revenue into an investment fund, how to manage the fund, and how to ensure the fund serving the objectives as they are designed. At each stage, these questions and others must be considered within the broader contexts of political and economic development of the country concerned. A central issue in the management of large resource revenue flows is that of transparency, which is becoming an increasingly challenging issue as NRFs try to compete in the primarily private global financial market. These studies provide the glimmers of light citizens need to understand how *their* funds are being used on *their* behalf.

References

Adams J (1996) Principals and agents. American Sociological Review 61(1):12–28

Al-Hassan A, Papaioannou M, Skancke M, Sung CC (2013) Sovereign wealth funds: Aspects of governance structures and investment management. IMF Working Paper, WP/13/231

Alsweilem KA (2015) A stable and efficient fiscal framework for Saudi Arabia. Belfer Centre for Science and International Affairs, Harvard Kennedy School

Alsweilem KA, Cummine A, Rietveld M, Tweedie K (2015) Sovereign investor models: Institutions and policies in managing sovereign wealth. Centre for International Development, Harvard Kennedy School

Amusan L, Saka L, Omede AJ (2017) Sovereign wealth fund and fiscal federalism in Nigeria (2011–14). Regional and Federal Studies 27(4):441–463

Arowolo D (2011) Fiscal federalism in Nigeria. Journal of Social Sciences 2(2):1–21

Balding C (2012) Sovereign wealth funds: The new intersection of money and politics. Oxford UP, Oxford

Bauer A (2014) Fiscal rules for national resource funds. Policy Brief, Natural Resource Governance Institute

Bauer A, Mihalyi D (2018) Premature funds. Natural Resource Governance Institute, Briefing

Bernstein S, Lerner J, Schoal A (2013) The investment strategies of sovereign wealth funds. Journal of Economic Perspectives 27(2):219–238

Bova E, Carcenac N, Guerguil M (2014) Fiscal rules and the procyclicality of fiscal policy in the developing world. IMF Working Paper, WP 14/122

Burnett WC, Lee AIN (1980) The phosphate supply system in the Pacific Region. GeoJournal 4(5):423–435
Caldecott B, Harnett E (2019) One planet sovereign wealth funds: Turning ambition into action. Oxford Sustainable Finance Programme Briefing Paper, University of Oxford
Capapé J, Schena PJ, Rose P, López D (2019) Sovereign wealth funds 2019: Managing continuity, embracing change. IE Centre for the Governance of Change
Chatham House (2014) Africa's sovereign wealth funds: Demand, development, and delivery. 5 Sep
Chilean Ministry of Finance (2017) Sovereign Wealth Funds Annual Report 2017. DIPRES. https://www.hacienda.cl/english/sovereign-wealth-funds/annual-report/annual-report-sovereign-wealth-funds-355125.html
Clark G, Dixon A, Monk A (2013) Sovereign wealth funds: Legitimacy, governance and global power. Princeton UP, Princeton
Clark GL, Monk AHB (2013) Principles and policies for in-house asset management. Journal of Financial Perspective 1(3):39–47
Cumming DJ, Wood G, Filatochev I, Reinecke J (eds) (2016) The Oxford handbook of sovereign wealth funds. Oxford UP, Oxford
Davis J, Ossowski R, Daniel J, Barnett S (2001) Stabilization and savings funds for non-renewable resources. Occasional Paper 205. IMF, Washington, DC
Dixon A, Monk A (2014) Frontier finance. Annals of Association of Am Geographers 104(4):852–868
Dolfsma W, Grosman A (2019) State capitalism revisited. Journal of Economic Issues 53(2):579–586
Edwards JB (2014) Phosphate mining and the relocation of the Banabans to Northern Fiji in 1945: Lessons for climate change-forced displacement. Journal de la société des océanistes 138–139:121–136
Erbil N (2011) Is fiscal policy procyclical in developing oil-producing countries? IMF Working Paper, WP/11/171
European Bank for Reconstruction and Development (2018) The fiscal implications for Kazakhstan of worldwide transition to a greener global economy. EBRD, London
Fasano U (2000) Review of the experience with oil stabilization and savings funds in selected countries. Working Paper WP/00/112. IMF, Washington, DC
Gale SJ (2016) The mined-out phosphate lands of Nauru, Equatorial Western Pacific. Australian Journal of Earth Sciences 63(3):333–347
Gelb A et al. (1988) Oil windfalls: Blessing or curse? Oxford UP, Oxford
Gelb A et al. (2014) Sovereign wealth funds and long-term development finance. World Bank Policy Research Working Paper 6776

Gelpern A (2011) Sovereignty, accountability, and the wealth fund governance conundrum. Asian Journal of International Law 1(2):289–329
Grand S, Wolff K (2020) Assessing Saudi vision 2030. Atlantic Council, June 2020
Hannesson R (2001) Investing for sustainability: The management of mineral wealth. Kluwer, Boston
Hart O (1995) Corporate governance. Economic Journal 105(430):678–689
Hood AN (2005) Australia adrift: The Timor Sea oil and gas dispute. Brown Journal of World Affairs 12(1):239–254
IMF (2004) Democratic Republic of Timor-Leste: 2004 Article IV consultation: Staff report. IMF Country Report NO. 04/321
IMF (2019) Mongolia: Staff report for the 2019 Article IV consultation
Johan SA, Knill A, Mauck N (2012) Determinants of SWF investment in private v public equity. Journal of International Business Studies 44(2):155–172
Katz M, Bartsch U, Malothra H, Cuc M (2004) Lifting the oil curse: Improving petroleum revenue management in Sub-Saharan Africa. International Monetary Fund, Washington, DC
La Porta R, Lopez-de-Silanes F, Shleifer A, Vishny RW (1998) Law and finance. Journal of Political Economy 106(6):1113–1155
Lie E (2018) Learning by failing. Scandinavian Journal of History 43(2):284–299
McCubbins MD (1985) The legislative design of regulatory structure. American Journal of Political Science 29:721–748
Miller GJ (2005) The political evolution of principal-agent models. Annual Review of Political Science 8:203–225
Monk A, Sharma R, Sinclair DLS (2017) Reframing finance. Stanford UP, Stanford
Moses J, Letnes B (2017) Managing resource abundance and wealth. Oxford UP, Oxford
OECD (2016) Corruption in the extractive value chain, OECD, Paris
Petersen C, Budina N (2003) Governance framework of oil funds: The case of Azerbaijan and Kazakhstan. World Bank, Washington, DC
Pratt L, Richards J (1979) Prairie capitalism. McClelland and Stewart, Toronto
Rashad M (2020) Saudi Arabia's PIF gets $40 billion from foreign reserves to fund overseas investments. Reuters, 30 May
Rietveld M, Bauer A (2014) Rule-based investment for natural resource funds. Policy Brief, Natural Resource Governance Institute
Rogoff K (1990) Equilibrium political budget cycles. American Economic Review 80:21–36
Roll S (2019) A sovereign wealth fund for the prince. German Institute for International and Security Affairs, SWP Research Paper 8

Salami A (2011) Taxation, revenue allocation and fiscal federalism in Nigeria. Economic Annals 56(189):27–50.
Sauvant KP, Sachs LE, Jongbloed WPFS (eds) (2012) Sovereign investment: Concerns and policy reactions. Oxford UP, Oxford
Scheiner C (2006) Drilling East Timor: Australia's oil grab in the Timor Sea. Multinational Monitor 27(1):30–32
Scheiner C (2015) Can the Petroleum Fund exorcise the resource curse from Timor-Leste? In: Ingram S, Kant L, McWilliam A (eds) A new era: Timor-Leste after the UN. Australian UP, Canberra, p 73–101
Smith PJ (1991) The politics of plenty: Investing natural resources in Alberta and Alaska. Canadian Public Policy 17(2):139–154
Smith B (2004) Oil wealth and regime survival in the developing world, 1960–1999. American Journal of Political Science 48:232–246
Stevens P (2006) "Resource curse" and how to avoid it. Journal of Energy and Development 31(1):1–20
Stevenson G (ed) (1981) Western separatism. Hurtig, Edmonton
Stone SE, Truman E (2016) Uneven progress on SWF transparency and accountability. PIIE
Thatcher M, Vlandas T (2016) Overseas state outsiders as new sources of patient capital. Socio-Economic Review 14(4):647–668
Tranøy BS (2010) Norway – The Accidental Role Model. In: Xu Y, Bahgat G (eds), The Political Economy of Sovereign Wealth Funds. Palgrave Macmillan, Hampshire, UK
Truman E (2010) Sovereign wealth funds: Threat or salvation? Peterson Institute for International Economies, Washington, DC
Truman E (2017) SWF transparency and accountability explored. Wake Forest Law Review 52:997–1026
United Nations (2017) Financing for development: Progress and prospects. United Nations, New York
Weingast BR, Moran MJ (1983) Bureaucratic discretion or congressional control. Journal of Political Economy 91(51):765–800
Williamson O (2000) The new institutional economics. Journal of Economic Literature 38(3):595–613
Wood G, Wright M (2015) Corporations and new statism. Academy of Management Perspectives 29(2):271–286
Wright C (2019) Asia sovereign wealth funds: Lifting the lid on Singapore's GIC. Euromoney, 9 May
Xu YC (2017) The Australian future fund. In: Cumming D et al. (eds) The Oxford handbook of sovereign wealth funds. Oxford UP, Oxford, p 622–650
Zider B (1998) How venture capital works Harvard Business Review, November/December

Correction to: The Political Economy of Natural Resource Funds

Eyene Okpanachi and Reeta Chowdhari Tremblay

Correction to:
E. Okpanachi and R. Chowdhari Tremblay (eds.),
***The Political Economy of Natural Resource Funds*, International Political Economy Series,**
https://doi.org/10.1007/978-3-030-78251-1

The original version of this book was inadvertently published with few errors and the same has been corrected.

The updated version of this book can be found at
https://doi.org/10.1007/978-3-030-78251-1_5
https://doi.org/10.1007/978-3-030-78251-1_7
https://doi.org/10.1007/978-3-030-78251-1_10
https://doi.org/10.1007/978-3-030-78251-1

E. Okpanachi and R. Chowdhari Tremblay (eds.), *The Political Economy of Natural Resource Funds*, International Political Economy Series,
https://doi.org/10.1007/978-3-030-78251-1_12

Index

E. Okpanachi and R. Chowdhari Tremblay (eds.), *The Political Economy of Natural Resource Funds*, International Political Economy Series,
https://doi.org/10.1007/978-3-030-78251-1

J

K

L

M

N

O

S